Advanced Distributed Measuring Systems — Exhibits of Application

River Publishers Series of Information Science and Technology

Series Editor

KC Chen
National Taiwan University
Taipei, Taiwan

Information science and technology ushers 21st century into an Internet and multimedia era. Multimedia means the theory and application of filtering, coding, estimating, analyzing, detecting and recognizing, synthesizing, classifying, recording, and reproducing signals by digital and/or analog devices or techniques, while the scope of "signal" includes audio, video, speech, image, musical, multimedia, data/content, geophysical, sonar/radar, bio/medical, sensation, etc. Networking suggests transportation of such multimedia contents among nodes in communication and/or computer networks, to facilitate the ultimate Internet. Theory, technologies, protocols and standards, applications/services, practice and implementation of wired/wireless networking are all within the scope of this series. Based on network and communication science, we further extend the scope for 21st century life through the knowledge in robotics, machine learning, cognitive science, pattern recognition, quantum/biological/molecular computation and information processing, biology, ecology, social science and economics, user behaviors and interface, and applications to health and society advance.

- Communication/Computer Networking Technologies and Applications
- Queuing Theory, Optimization, Operation Research, Stochastic Processes, Information Theory, Statistics, and Applications
- Multimedia/Speech/Video Processing, Theory and Applications of Signal Processing
- Computation and Information Processing, Machine Intelligence, Cognitive Science, Decision, and Brian Science
- Network Science and Applications to Biology, Ecology, Social and Economic Science, and e-Commerce

For a list of other books in this series, please visit www.riverpublishers.com

Advanced Distributed Measuring Systems — Exhibits of Application

Prof. Vladimír Haasz

Czech Technical University in Prague, CZ

Published 2012 by River Publishers
River Publishers
Alsbjergvej 10, 9260 Gistrup, Denmark
www.riverpublishers.com

Distributed exclusively by Routledge
4 Park Square, Milton Park, Abingdon, Oxon OX14 4RN
605 Third Avenue, New York, NY 10158

First published in paperback 2024

Advanced Distributed Measuring Systems — Exhibits of Application / by Vladimír Haasz.

Routledge is an imprint of the Taylor & Francis Group, an informa business

Publisher's Note
The publisher has gone to great lengths to ensure the quality of this reprint but points out that some imperfections in the original copies may be apparent.

ISBN: 978-87-92329-72-1 (hbk)
ISBN: 978-87-7004-526-1 (pbk)
ISBN: 978-1-003-48234-5 (ebk)

DOI: 10.1201/9781003482345

Editor's Biography

Vladimir Haasz finished the Czech Technical University (CTU) in Prague, Faculty of Electrical Engineering (FEE) in 1972 (branch "Cybernetics"). Since that year he has been with Department of Measurement. He got his Ph.D. degree in 1977, he defended his habilitation thesis in 1991, and he was named full professor of Measurement technology in 1999. He was active half a year at ETH Zurich in 1994. Now he manages the Department of Measurement at CTU-FEE.

Vladimir Haasz is the member of TC 12 "Quantities and Values" of the Czech Institute of Standardization, the member of IMEKO (International Measurement Confederation) TC-4 — Measurement of Electrical Quantities, and the honorary member of the Scientific Counsel of CTU.

He has been interested in the field of measuring systems of electrical quantities, sampling methods of measurement of non-harmonic waveforms, and in the last years especially in testing of dynamic quality of AD modules and their EMC. He gives lectures on the basic course "Electrical Measurement and Instrumentation" and on the optional course "Advanced Instrumentation".

Preface

The idea to publish a book concerning advanced distributed measuring systems has arisen based on discussions that passed off during preparation of the 6th IEEE International Conference on Intelligent Data Acquisition and Advanced Computing (IDAACS'2011). The increasing interest in this area documented by accretion of sent papers as well as an offer to publish the selected re-written and expanded papers as an edited volume by River Publishers created the decision to try to realise it.

The most interesting papers concerning embedded applications, small distributed systems including automotive systems, monitoring systems based on wireless networks, problems of synchronisation in large DAQ systems and also virtual instrumentation were selected and their authors were addressed to re-write the paper into the form of book chapter. The chapters are conceived more widely than the original papers, they include not only system design solution in wider context but also relevant theoretical parts, achieved results and possible future ways of design and development of similar systems.

The dissemination of new approaches in the area of distributed measuring systems and presentation of progressive up-to-date solutions of their applications is the main goal of the book. Its contents could be convenient for personnel of firms deal with control systems, automotive electronics, airspace instrumentation, health care technology etc. as well as for academic staff and postgraduate students in electrical, control and computer engineering. The editor of the book as well as the authors of the individual chapters hope that readers will find there both the new knowledge from up-to-date solutions in the area of distributed measuring systems and inspiration for solving their problems in this area.

At first, I would like to thank the authors of all chapters for their work that they carried out by chapter preparation. Thanks go also to prof. Anatoly Sachenko, IDAACS Conference Co-Chairmen and my colleagues from IDAACS Conference International Advisory Board, and at last but not least to Mr. Rajeev Prasad, who helpfully and patiently enabled to realize this project.

Prague, January 2012

Vladimir Haasz
Editor

Contents

1

Introduction

Vladimir Haasz

Czech Technical University in Prague, Faculty of Electrical Engineering

Measuring systems are an essential part of all automated production systems. They also serve to ensure quality of production or they help to assure the reliability and safety in various areas. The same applies in principle likewise for fields of telecommunication, energy production and distribution, health care etc. Similarly no serious scientific research in the field of natural and technical sciences can be performed without objective data about the investigated object, which is usually acquired using measuring system. Demands on the speed and accuracy of measurement increase in all areas in general. These are the grounds for publishing this book.

The book offers 8 examples of typical laboratory, industrial and biomedical applications of advanced measuring and information systems including virtual instrumentation. It arose based on the most interesting papers from this area published at IDAACS'2011 conference. They were selected to cover the systems from embedded applications across small distributed systems used in cars to large distributed monitoring systems based on wireless networks used e.g., in health care technologies. Progressive up-to-date solutions are presented in the following examples. However, single chapters include not only system design solution in wider context but also relevant theoretical parts, achieved results and possible future ways of design and development.

Advanced Distributed Measuring Systems — Exhibits of Application, 1–4.

At the beginning there are presented the service-oriented distributed measurement and control systems. The chapter deals with a service-oriented middleware platform targeted on instrumentation, called Service eXtensions for Instrumentation (SXI). The middleware platform is used to build a Distributed Measurement and Control System (DMCS) around a physical process equipped with all the instrumentation needed to run control loops for pressure, level, flow and temperature, quantities widely found in the industrial process. The tests have been made to evaluate the performance of the system in terms of sampling frequency, communication delays and behaviour of control loops. The methodology of each experiment is described, results are analysed and conclusions are extracted.

The wireless sensor networks (WSNs) have been increasingly used in the past years. Their design and development imply software simulations and hardware tests. The possible solution of this task is described in the following chapter. It depicts the testing environment developed for WSN nodes energy consumption monitoring and hardware simulation of the nodes behaviour in a small scaled network. It is designed to control individually the sensed data and the power supply of the nodes from the network using a wired network of microcontroller based modules that are assigned to one or several sensors. These modules are linked with the same CAN bus as a computer system used to manage both networks (WSN and CAN).

A number of practical applications demand very fast data acquisition and real time processing. The up-to-date solution using embedded application is described in the 4th chapter in connection with the following application. On-line ultrasonic monitoring of aqueous solutions involves acquisition of an ultrasonic waveform either propagated through a solution or reflected from a reflector placed inside it. Most conventional ultrasonic instruments upload the waveform to a separate computer, which calculates from this waveform e.g., the propagation delay and the attenuation coefficient. This approach limits the achievable update rate and it requires use of a computer in addition to the ultrasonic instrument itself. The presented design utilizes embedded processing of the acquired waveforms in the ultrasonic instrument using a softcore MicroBlaze processor as a part of the overall FPGA design. A dedicated averaging subsystem (used to reduce noise) and an interleaved sampling subsystem (allowing to achieve the equivalent sampling frequency of 1 GHz using a 50 MHz-clocked ADC) operate alongside each other, and are controlled by the MicroBlaze. After the completion of the waveform acquisition the MicroBlaze implements a zero crossing algorithm to extract the sought propagation delay. The instrument was applied for monitoring of neutralization reactions with the update rate of 5 measurements (i.e., acquiring and processing a full interleaved waveform and after it reporting the ultrasound propagation time) per second.

The other problem in some large DAQ systems, which is necessary to be solved, can be synchronization of data acquisition. It is important for recording of dynamic actions and data acquisition of physical quantities over a wide geographical area. The chapter deals with synchronization of distributed measurement systems, especially synchronization using PTP (Precise Time Protocol, IEEE 1588). A practical case study presents a Master Clock module synchronized by GPS receiver. This module serves as high quality time base for PTP based system.

Quite different specific demands are put to systems used in automotive area. Special interfaces are used here regarding reliability and safety. FlexRay is an incoming automotive distributed system standard for safety critical applications like x-by-wire. The first part of the chapter provides an introduction to the FlexRay standard, presents a model of FlexRay. Synchronization mechanism, validation of this model and the offers of its usage for measurement of parameters of synchronization mechanism in real FlexRay networks is presented. The second part presents a case study of FlexRay application in a distributed implementation of ABS (Anti-lock Braking System) in vehicles.

The measuring systems including virtual instrumentation are also often used in the monitoring of environment. Different types of gas sensors are used to implement applications, which investigate ambient air pollution levels. Carbon monoxide (CO) concentration measurement is an integral part of dedicated environment monitoring stand-alone specialized systems. An alternative solution for CO concentration monitoring, based on an original virtual instrumentation concept, is presented here. The advantages of the proposed application include data logging, statistical calculations, remote access or software and hardware flexibility. Comparative experimental results are also provided.

The last two chapters are related to applications in health care technologies. The first one deals with using of up-to-date MEMS (micro electromechanical systems) in such application. Advancements in MEMS have produced IMU (Inertia Measurement Unit) sensors that can measure precisely the motion parameters that traditionally have been calculated as derivatives of displacement measured parameters. These sensors have the potential to cover the majority of human movements. An implementation of such a system using two pairs of 3D accelerometers and gyroscopes and determining the exact position of a joint by measuring the relative position of the two adjoined segments is presented here. Such a system can provide joint kinematic information overcoming shortcomings of video based motion analysis systems.

The second one presents continuous real-time monitoring of assisted livings through wireless body sensor networks. It proposes the wireless body sensor networks (WBSNs) as an enabling technology for a rich variety of application domains, from e-Health to e-Factory. In particular, the chapter describes reference network architectures, effective programming frameworks and novel applications in important application domains for WBSNs. Finally, a WBSN-based system in the e-Health application domain concerning all possible remote monitoring of assisted livings is described and analysed.

2

Service eXtensions for Instrumentation (SXI)

Vítor Viegas[1,2], P. Silva Girão[2,3], Miguel Pereira[1,2]

[1]*ESTSetúbal/LabIM, Instituto Politécnico de Setúbal, Setúbal, Portugal, vitor.viegas@estsetubal.ips.pt*
[2]*Instituto de Telecomunicações, Lisboa, Portugal*
[3]*DEEC/IST, Universidade Técnica de Lisboa, Lisboa, Portugal*

2.1 Aim

The Chapter presents a service-oriented middleware platform that takes advantage of the communication model provided by WCF services and the information model defined by the IEEE 1451.1 clause. The proposed platform was used to control a physical process equipped with all the instrumentation needed to run control loops for pressure, level, flow and temperature, quantities widely found in the process industry. Tests were made to evaluate the performance of the system in terms of sampling frequency, communication delays and behavior of control loops. The methodology of each experiment is described, results are analyzed and conclusions are extracted.

2.2 Introduction

Over the years, the evolution of measurement and control systems has been done by incorporating advances in other fields of knowledge, in particular the fields of microelectronics and information technologies (see Figure 2.1).

Advanced Distributed Measuring Systems — Exhibits of Application, 5–33.

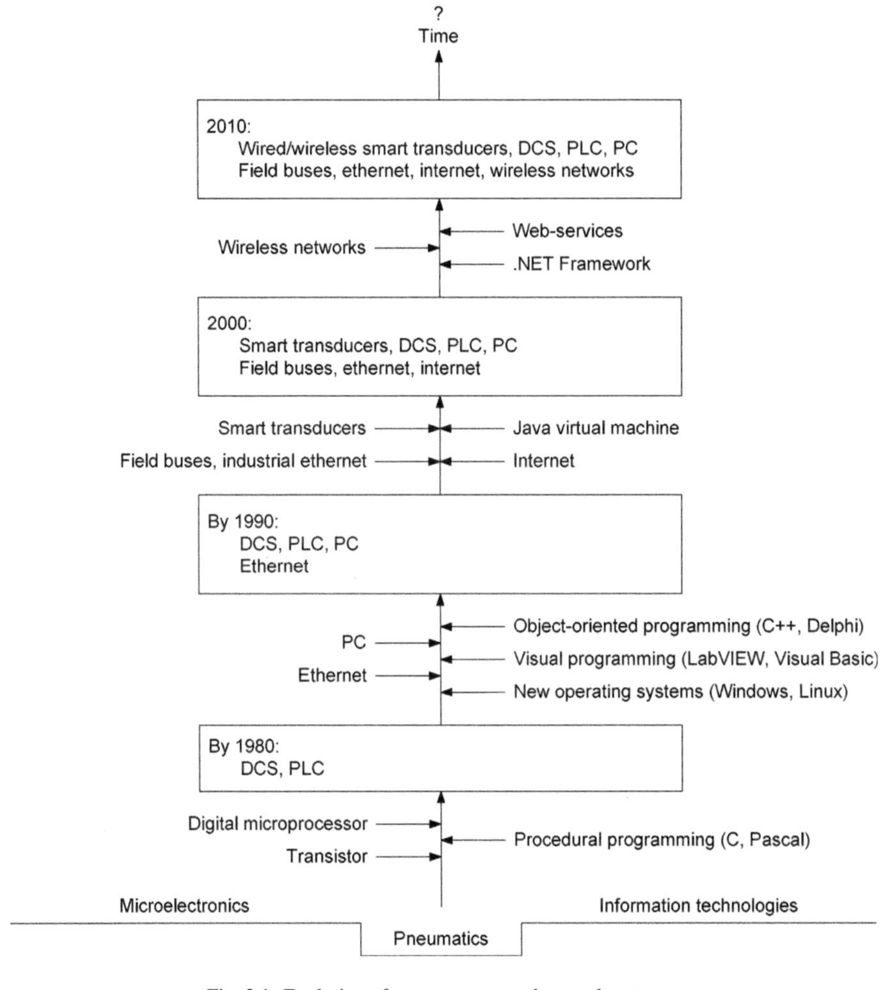

Fig. 2.1 Evolution of measurement and control systems.

The introduction of the digital microprocessor drove to the replacement of old pneumatic controllers by modern Programmable Logic Controllers (PLC) and Distributed Control Systems (DCS). Since then, the Personal Computer (PC) gained its place in non-critical tasks (such as configuration, supervision and data analysis) and the ethernet became the preferred network for data communications. The increasing performance of the microprocessor (in terms of processing power, size, consumption and price) enabled its installation near to the primary element of measurement/actuation, all embedded in the same package, leading to the concept of "smart transducer". The transducers became smart enough to execute routines of self-identification, self-calibration and self-diagnostics, as well as to communicate with each other through a wired or wireless fieldbus.

In the field of information technologies, new development frameworks were released to address the demand for more and better software. These frameworks integrate all the tools needed to implement and execute applications, such as Object-Oriented (OO) languages (more productive than traditional procedural languages), visual editors (like those available in Visual Basic and LabVIEW), pre-built code libraries, intuitive debugging tools and high-performance virtual machines. A good example is the .NET Framework [1, 2], released in 2002 by Microsoft for Windows operating systems. The .NET Framework consists of three main components:

- **Programming languages**: The .NET Framework includes two last-generation programming languages: C#, which borrows heavily from C and C++ but adds high-level facilities such as automatic memory management, security and threading; and VB.NET, which is the evolution of Visual Basic. The high-level code is compiled into an intermediate language known as "Microsoft Common Intermediate Language" (MSIL).
- **Common Language Runtime (CLR)**: Virtual machine that takes the MSIL code and compiles it to native processor instructions. The CLR makes .NET applications platform-independent because it provides an abstract execution environment that is separated from the underlying processor and operating system.
- **Class libraries**: The .NET Framework provides a vast set of dynamically loadable libraries that can be used by the programmer to perform common tasks. Hundreds of pre-built classes are available to construct graphical user interfaces, access files, interact with databases, handle network communications, and so on. Correlated classes are grouped in software packages according to their functional affinities.

The gains in terms of software productivity were followed by the development of cross-platform web-service-oriented applications [3, 4]. These applications cooperate in heterogeneous environments (like the internet) by calling remote methods (services) between them. Interoperability is achieved by imposing standards that describe the behavior of the service and the way to access it, regardless of its underlying implementation. This idea is not new, but new is the fact that web services are getting a wide acceptance among the software community. Being supported by all the major software companies around the world (such as IBM, Microsoft and Sun Microsystems), web services have the chance to become the first widely used middleware solution and the answer for many software interoperability problems.

The .NET Framework provides full support for service-oriented applications through a software package called "Windows Communication Foundation" (WCF) [5, 6, 7]. The WCF contains pre-built classes to implement "standard" web services (intended for communications between cross-platform applications on the internet), as well as "extended" services (intended for communications between uni-platform applications, inside the same machine or in different machines, on the intranet or internet). The rich set of functionalities provided by WCF services (much richer than that provided by web services), makes them the reference model for communications in Windows environments.

The advantages provided by the .NET Framework (in terms of software productivity and interoperability) can be complemented by those provided by the IEEE 1451 standard [8, 9], an industry initiative to simplify transducer connectivity to computers and communication networks. Interoperability is achieved by adopting standard hardware and software interfaces that act as "plugs" where heterogeneous components can connect and work together. These interfaces are specified in seven clauses:

- **Clause IEEE 1451.0 [10]**: Introduces the concept of "Transducer Interface Module" (TIM) as being a peripheral that connects transducers to the computer. The module can be seen as a Data AcQuisition (DAQ) board that contains Analog-to-Digital (AD) and Digital-to-Analog (DA) converters, as well as data structures for self-identification, self-calibration and self-diagnostics. The clause defines the common functions of the TIM, high-level communication protocols to communicate with it, and Transducer Electronic Data Sheet (TEDS) structures to describe it.

- **Clause IEEE 1451.2 [11]**: Details the concept of TIM for the case when it connects to the computer through a point-to-point wired interface.
- **Clause IEEE 1451.3 [12]**: Details the concept of TIM for the case when it connects to the computer through a multipoint wired interface.
- **Clause IEEE 1451.4 [13]**: Defines digital communication circuits and TEDS structures to be added to analog transducers in order to make them compatible with the IEEE 1451 standard.
- **Clause IEEE 1451.5 [14]**: Details the concept of TIM for the case when it connects to the computer through a wireless interface.
- **Clause IEEE 1451.7 [15]**: Defines data structures and commands to communicate with sensors attached to a Radio Frequency IDentification (RFID) interface.
- **Clause IEEE 1451.1 [16, 17]**: Defines a generic information model capable of representing the functionalities of any networked transducer. The model is composed by a hierarchy of classes divided in three main categories:
  - **Blocks**: Classes intended to acquire and process data. Three types of blocks are defined: (i) the processor block, which represents the application as whole and provides common resources to other objects; (ii) transducer blocks, which provide high-level functions to interact with transducers (such as read sensors, write actuators, read status registers, read/write interrupt masks and read/write TEDS structures); and (iii) function blocks, which implement data processing algorithms (such as math operations, multiplexers/demultiplexers, time windows, digital filters, Proportional-Integral-Derivative (PID) controllers, and so on).
  - **Components**: Classes intended to encapsulate data. Two types of components are defined: (i) parameters, which are used to store volatile data (such as field variables, setpoints and computations); and (ii) files, which are used to store persistent data (such as TEDS structures).

— **Services**: Classes that implement two communication models: (i) the client/server model for one-to-one communications; and (ii) the publish/subscribe model for one-to-many communications.

Together, the .NET Framework and the IEEE 1451 standard can be a major contribution towards open and interoperable measurement and control systems. Following this idea, we present a service-oriented middleware platform — named "Service eXtensions for Instrumentation" (SXI) — that takes advantage of the communication model provided by WCF services and the information model defined by the IEEE 1451.1 clause. This approach is very promising but has some pitfalls and drawbacks, in particular those related with the overhead introduced by additional software layers involved in data transfer and processing. To clarify these performance issues, we took the SXI platform and used it to control a physical process. Tests were made to evaluate the performance of the system in terms of sampling frequency, communication delays and behavior of control loops. The results obtained can serve as guidelines and benchmarks for the future.

The Chapter is organized as follows: Section 2.3 explains the architecture of the measurement and control system; Section 2.4 presents the physical process used as test bench; Section 2.5 describes the tests performed; and Section 2.6 extracts conclusions.

2.3 System Architecture

The system is based on the SXI platform as it provides the "bricks" with which we can build distributed control and supervision applications.

2.3.1 Service-Oriented Middleware Platform

The SXI platform merges the strengths of the IEEE 1451.1 clause with the benefits provided by the .NET Framework and the WCF. The idea was to materialize the 1451.1-information model using last-generation software technologies in order to obtain an open middleware solution targeted for instrumentation.

The SXI platform implements a fully-functional subset of the 1451.1-information model (see Table 2.1). The subset is composed by blocks, components and services, making a total of 19 classes, all coded in VB.NET (version 2008) and assembled in the reusable library *sxi.dll*. Despite all the attempts

Table 2.1. Information model of the SXI platform.

Class name	Description
Root	It is the base class for all other classes.
Entity	It provides functionalities to identify and localize objects in the context of the application or across the network.
Block	It is the base class for all blocks. It provides functionalities to retrieve information about the block itself, to change its executing state, and to interact with its owned objects.
PBlock	Processor block that represents the application as a whole. It centralizes information about the application and provides common resources to other objects.
FBlock	It is the base class for all function blocks. It provides basic functionalities to execute processing algorithms.
HysteresisFBlock	It implements the Schmitt-trigger algorithm widely used in on/off control.
PIDFBlock	It implements the PID algorithm widely used in process control.
TBlock	It is the base class for all transducer blocks.
DAQmxTBlock	Transducer block that works with DAQ boards compliant with the *DAQmx* driver from National Instruments.
Component	It is the base class for all components.
Parameter	It represents a network-visible variable that can be read and written.
ParameterWithUpdate	It provides functionalities to synchronize the value of the parameter with its owning block.
PhysicalParameter	It adds metadata structures to describe the contents of the parameter.
ScalarParameter	Parameter that represents a physical quantity quantified by a mathematical scalar.
DAQmxTChannel	Component that exposes the properties of a DAQ channel.
Service	It is the base class for all services.
Client	A kind of improved WCF proxy.
Publisher	It issues publications on a UDP multicast address.
Subscriber	It listens to publications on a UDP multicast address.

Notes:
1) Classes listed in bold are non-abstract.
2) The indexing of each class represents its position in the hierarchy.

to follow as closely as possible the 1451.1-information model, some changes had to me made, such as the modification of some classes and the introduction of new ones, the redefinition of some data types and methods, and the use of native WCF proxies instead of client and publisher ports. Therefore, we do not expect the SXI platform to be compatible with the IEEE 1451.1 clause, but rather an alternative (and improved) approach.

All non-abstract classes of the SXI platform are implemented as WCF services marked with the following attributes:

- ***InstanceManagement*** = **single**: This means that the service is a unique instance that is shared by all the clients.

- ***ConcurrencyMode*** = **single**: This means that remote calls are served one at a time, in absolute exclusivity. As a result, the internal state of the service is preserved between remote calls.
- ***UseSynchronizationContext*** = **false**: This means that each remote call is served by a dedicated thread. In each moment, there may be multiple threads servicing multiple remote calls on different services.

All WCF services natively support the client/server communication model. Whenever a service is created, it registers itself on a client/server endpoint and exposes its methods on the network. If a client wants to invoke a method, it gets the dispatch address of the service, creates a proxy at run-time, invokes the remote call, and collects the results (if any). Client/server endpoints can use one of the following bindings:

- ***BasicHttpBinding***: This binding is totally compatible with web services. It uses the Simple Object Application Protocol (SOAP) [18] to format messages and the Hyper Text Transport Protocol (HTTP) to transport them over the wire. It promotes interoperability over performance, making it suitable for communications between cross-platform applications in the internet.
- ***WSHttpBinding***: This binding also uses the SOAP format and the HTTP transport. In addition, it supports *WS** extensions [19] making it more versatile but also less interoperable. It is indicated for communications between uni-platform applications in the internet.
- ***NetTcpBinding***: This binding uses a proprietary binary protocol to format messages and the Transmission Control Protocol (TCP) to transport them over the wire. It promotes performance over interoperability, making it suitable for communications between uni-platform applications inside an intranet.

By choosing the right binding, client/server communications can be tuned for performance or interoperability depending on the application needs.

The publish/subscribe communication model is trickier as it requires two dedicated classes to implement it (the classes *Publisher* and *Subscriber*). These classes use a non-standard binding [20], based on the User Datagram Protocol (UDP), to issue/intercept publications to/from a multicast address.

A publication is a method with no return values (equivalent to a "one-way" message) that is called by the publisher and forwarded to all subscribers registered in the multicast address.

On the field side, the SXI platform works with DAQ boards compliant with the *DAQmx* driver [21] from National Instruments (NI). The interface is done by the class *DAQmxTBlock*, which wraps the functions contained in the driver that "indeed" communicate with the board. The life-cycle of a *DAQmxTBlock* is as follows:

1. At design-time, using the Measurement and Automation eXplorer (MAX), a software tool provided by NI, the developer configures and saves tasks involving one or more channels. A task is a data structure that contains all the information needed to acquire/generate signals, including properties that describe the digitizer (such as hardware channels, number of samples, sampling frequency and trigger settings), and properties that describe the transducer itself (such as transducer type, range and units). A channel is in turn a hardware input/output that is part of a task. Each channel can be automatically configured by reading the TEDS of the attached transducer according to the directives of the IEEE 1451.4 clause. If the transducer has no embedded TEDS, the developer can associate it a virtual TEDS in the form of a *.ted* file. Detailed information about *DAQmx* tasks can be found in [22, 23, 24].
2. At run-time, the *DAQmxTBlock* is created, initialized and executed. During initialization, the *DAQmxTBlock* loads a pre-configured task and creates auxiliary objects to support it (one parameter for the whole task and one *DAQmxTChannel* for each channel). The parameter provides methods to read/write data from/to transducers, as well as metadata structures that describe the meaning of the data exchanged. The *DAQmxTChannel* provides information about the channel where the transducer is attached, including methods to retrieve the underlying TEDS.

At the present, the SXI platform is limited to one transducer block, two function blocks and parameters of type scalar. In the future, the information model shall include new transducer blocks (to improve field communications), new function blocks (to improve data processing capabilities), and new classes to support multidimensional parameters.

2.3.2 Control Stations

Control stations are object-oriented applications that use the SXI platform to control physical processes. They are hosted in computers connected to an intranet allowing the execution of local control loops (inside a control station), as well as distributed control loops (involving two or more control stations).

A control station resembles a PLC in the sense that it executes the control routine periodically by means of a timed loop. On every loop iteration, data is acquired from sensors (using one or more transducer blocks), control algorithms are executed (using one or more function blocks), and output values are written to actuators (using one or more transducer blocks), always by this order, as depicted in Figure 2.2. Each block owns a predefined set of

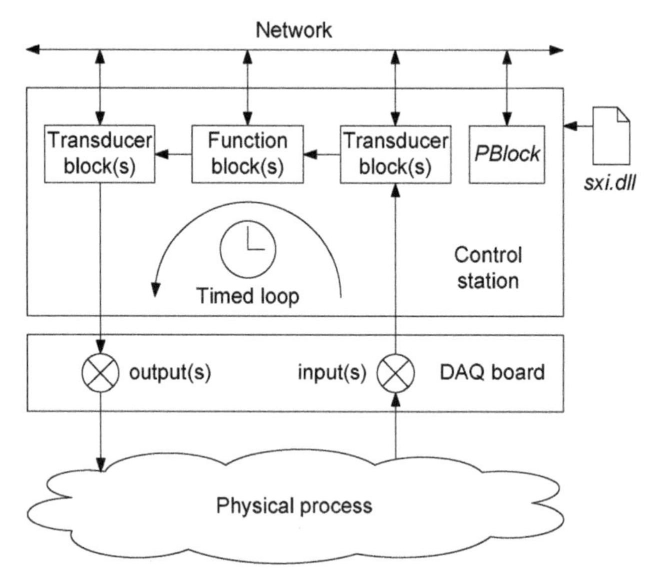

Fig. 2.2 Data flow inside a control station.

components and services, all working as its assistants: parameters are used to store block variables, publishers are used to broadcast block variables across the network, and subscribers are used to collect data from the network and feed block variables.

All control stations must provide publications to announce themselves to the network and to describe the WCF services they expose. These publications must be issued at start-up and on demand (and optionally at regular intervals as a heart beat).

Figure 2.3 proposes a Graphical User Interface (GUI) for control stations. The interface is dominated by a tree list where the operator can see all objects created by the application. Each object is identified by its name and is positioned according its owning relation. In the middle-column, the operator can see (and edit) the value of application variables.

2.3.3 Engineering Stations

Engineering stations are object-oriented applications that use the SXI platform to configure and monitor control stations. They are hosted in computers connected to the same intranet as that of the control stations. This arrangement

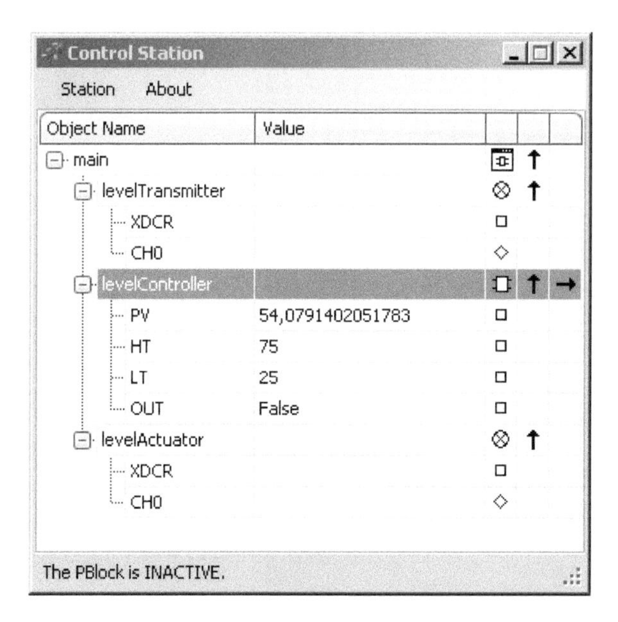

Fig. 2.3 GUI of control stations.

makes possible for an engineering station to "take care" of the control stations connected in the neighborhood. An engineering station serves two main purposes:

- **System configuration**: This task is done by intercepting publications carrying announcements of control stations and their WCF services. The attached data is extracted and used to build a virtual image of the network, including the type, name, identifier and dispatch address of all registered services. This information is all that is needed to create a proxy and invoke any method of any service on any station. In other words, it is all that is needed to configure the entire system.
- **Data monitoring**: This task is done by intercepting publications carrying block variables. The attached data, which includes the value of the variable and its identification, is extracted and optionally logged to a file.

Figure 2.4 presents the GUI of engineering stations. On the left pane, the operator can see all the services registered in the network. By clicking on a given service, the operator gains access to its properties on the right pane.

2.4 Physical Process

The physical process used as test bench is a training plant (model TE34 from Plint & Partners Ltd [25]) that includes all the instrumentation needed to run the following control loops (see Figure 2.5):

- **Pressure loop**: The pressure inside the closed tank C2 is measured by the transmitter PT and is controlled by operating the control valves PCV1 and PCV2 (which form a complementary pair). The pressure increases when PCV1 opens and PCV2 closes, and vice-versa.
- **Level loop**: The water level inside the closed tank C2 is measured by the transmitter LT and is controlled by operating the control valves FCV1 and FCV2 (which also form a complementary pair). The water is continuously pumped from the open tank C1 to the closed tank C2 and returns back through FCV1 and the hand valve HV.

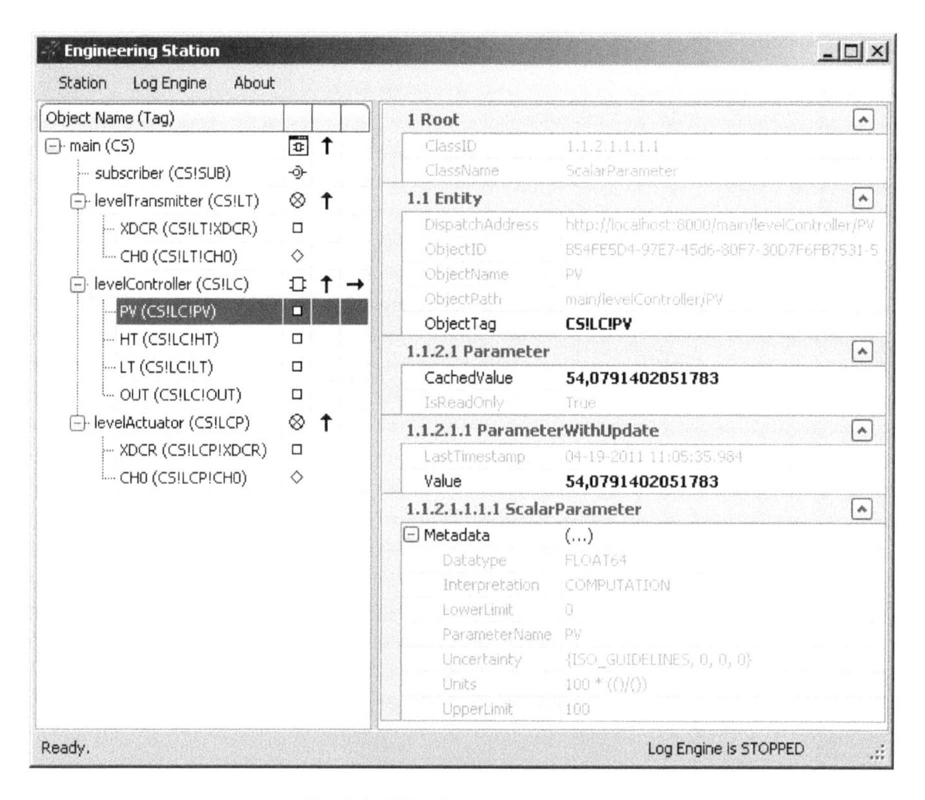

Fig. 2.4 GUI of engineering stations.

- **Temperature loop**: The temperature of the water entering in the open tank C1 is measured by the transmitter TT and is controlled by operating the control valve TCV. The flow of hot water is constant while the flow of cold water can be adjusted. The tank C1 is equipped with an overflow tube that drains water out into the sewer if it gets too high.

Table 2.2 lists all the instruments installed on the plant. The interface with pneumatic instruments was done using pressure-to-voltage converters (for the transmitters) and voltage-to-pressure converters (for the control valves). All converters, transmitters and control valves were properly verified and calibrated before experiments took place.

Two control stations were used to execute the control loops: the pressure and level loops were assigned to control station number one (CS1) and the

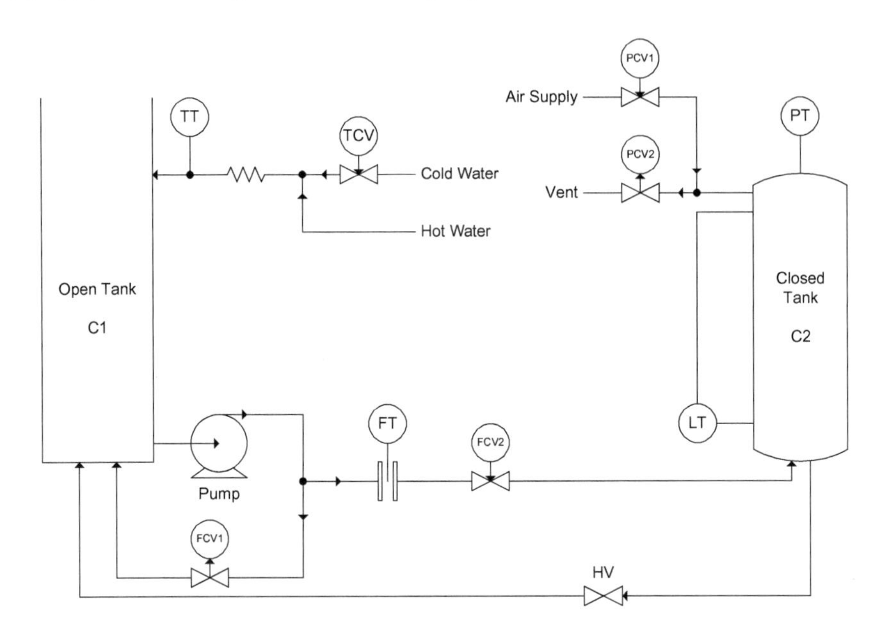

Fig. 2.5 Physical process.

Table 2.2. Field instruments.

Tag	Manufacturer	Reference	Signal	Summary
PT	FOXBORO	821GM-IS1NM1-A	4-20 mA	Pressure transmitter
PCV1	MASONEILAN	29000	0.2-1 bar	Pressure control valve
PCV2	MASONEILAN	29000	0.2-1 bar	Pressure control valve
LT	FOXBORO	15A-LS1-R	0.2-1 bar	Level transmitter
FCV1	MASONEILAN	29000	0.2-1 bar	Flow control valve
FCV2	MASONEILAN	29000	0.2-1 bar	Flow control valve
FT	FOXBORO	15A-LS1	0.2-1 bar	Orifice flowmeter
TT	FOXBORO	E94-P625	4-20 mA	Temperature transmitter
TCV	MASONEILAN	29000	0.2-1 bar	Temperature control valve

temperature loop was assigned to control station number two (CS2). Both control stations were equipped with a DAQ board model USB-6008 from NI.

The control strategy of station CS1 consists of two Proportional-Integral (PI) controllers implemented by two *PIDFBlocks* with derivative component equal to zero. The control strategy of station CS2 is slightly different as it consists of one fully-functional PID controller. In this case, the derivative component is needed to compensate the high inertia of the temperature loop.

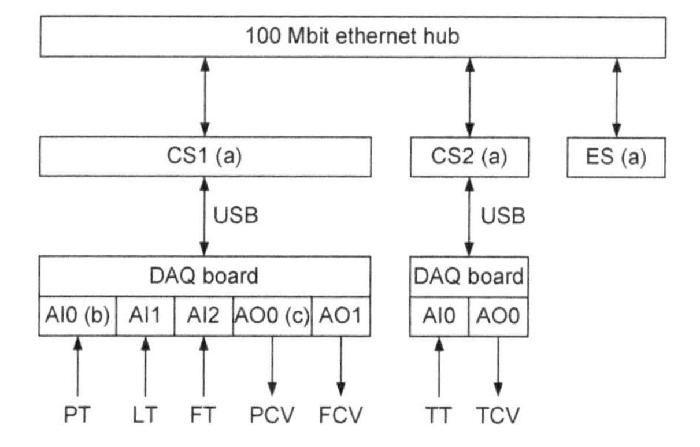

Notes:
a) Computer with the following characteristics: motherboard Intel DG41RQ Essential Series, processor Intel Pentium Dual Core E5300, 4 GB of RAM DDR2 800 MHz, hard drive Seagate 320 GB SATA II ST3320613AS, operating system Windows XP Home SP3, .NET Framework version 3.5.
b) Analog Input number zero.
c) Analog output number zero.

Fig. 2.6 Final arrangement of the system.

Finally, an engineering station (ES) was added to configure and monitor the system. The three stations were installed on three distinct machines, all providing the same computational support. The three computers were connected to an eight-port ethernet hub (model 3C16753 from 3Com) forming a 100 Mbit intranet (see Figure 2.6).

2.5 Experimental Results

The system was evaluated in terms of sampling frequency, communication delays and behavior of control loops.

2.5.1 Sampling Frequency

Each control station executes its control routine periodically by means of a timed loop. Precise timing is achieved by performing passive waits with a resolution of 1 ms. The elapsed time between loop iterations, here called "cycle time", determines the sampling frequency of the control station and has a strong impact on the quality of control algorithms.

The experiments to evaluate the sampling frequency of the system took place in station CS1 because its workload is larger than that of station CS2. The adopted methodology was as follows:

1. A *stopwatch* object was added to the code in order to measure the elapsed time between loop iterations. The *stopwatch* object is a diagnostic tool that is capable of measuring time intervals with a microsecond resolution [26].
2. Station CS1 was started assuming a nominal cycle time of 100 ms. About 1000 samples of effective cycle time were acquired and logged to a file.
3. Step 2 was repeated for the following nominal cycle times: 50 ms, 200 ms and 1000 ms. Figure 2.7 presents the distribution of the samples for 50 ms and 100 ms. Table 2.3 summarizes the results of all experiments.
4. If station CS1 remains in operation for a long period of time, the memory available in the computer gradually decreases, returning to the baseline from time to time. When this happens, the Hard

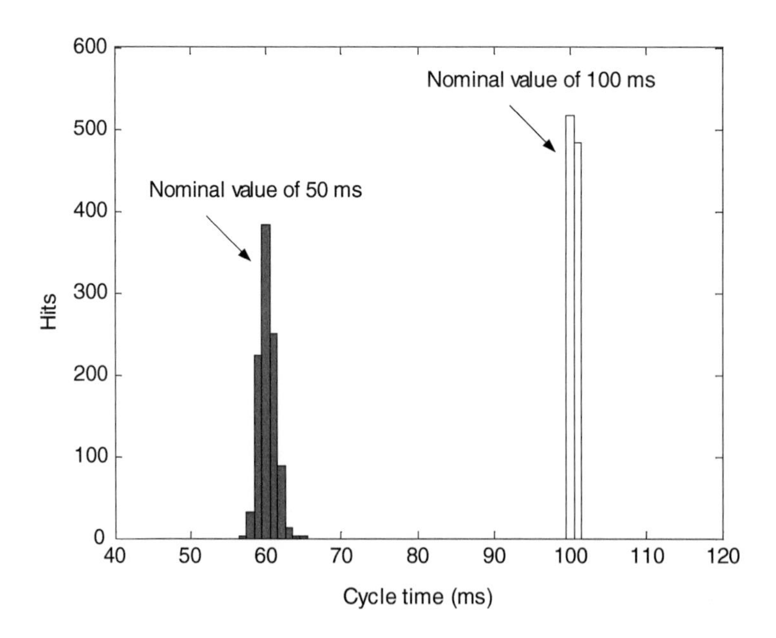

Fig. 2.7 Effective cycle time histogram for 50 ms and 100 ms.

Table 2.3. Cycle time of station CS1.

Nominal value (ms)	Effective value (ms)			Mean relative error (%) (a)	Centered samples (%) (b)
	Mean	Min	Max		
50	60.19	57	65	20.38	0
100	100.48	100	101	0.480	100
200	200.52	200	201	0.260	100
1000	1000.46	1000	1001	0.046	100

Notes:
a) Defined as 100 × |"nominal value" – "mean effective error" | /"nominal value".
b) Percentage of samples inside the interval "nominal value" ± 1 ms.

Disk Drive (HDD) is heavily accessed and the cycle time suffers variations of tens of milliseconds.

The collected data can be analyzed as follows:

- Above 100 ms inclusive, the mean value of effective cycle time is very close to the nominal value. The mean relative error tends to decrease suggesting that larger cycle times are more accurate, as expected.
- Above 100 ms inclusive, the percentage of centered samples is 100%, meaning that the Windows XP performs reasonably well although it is not a real-time operating system.
- Below 100 ms, the nominal cycle time is not satisfied. The minimum mean value of effective cycle time is approximately 60 ms, which corresponds to a maximum sampling frequency of 16 Hz. Given the inertia of the physical process, a sampling frequency of 2 Hz was chosen to control it.
- The strong variations observed in the cycle time (coincident with intense HDD activity) are caused by the garbage collector of the .NET Framework [27]. When it starts working, all applications are stopped waiting for it to clean up the memory.

2.5.2 Communication Delays

The WCF uses a chain of logic channels to assemble and disassemble the messages transmitted from the sender to the receiver(s). By changing the composition of the chain (adding, removing or reconfiguring logic channels), it is possible to refine the way messages are processed and shipped over the wire. This arrangement is very flexible but introduces delays in the communication path that degrade the responsiveness of the whole system.

2.5.2.1 Client/Server delay

The experiments to evaluate the delay of client/server communications took place in stations CS1 and CS2. The adopted methodology was as follows:

1. The sampling frequency of both control stations was adjusted to 2 Hz. The control routine of station CS1 was modified to periodically read the setpoint of the temperature loop running in station CS2. In this scenario, station CS1 is the client and station CS2 is the server.
2. Two instructions were added on the client side, just before and after the remote call, to generate a positive pulse on the parallel port data bits. This pulse delimits the read operation including the communication delays.
3. Two instructions were also added on the server side, at the beginning and end of the service, to generate a positive pulse on the parallel port data bits. This pulse delimits the read operation excluding the communication delays.
4. Both pulse signals were connected to a XOR gate to discriminate two delays (see Figure 2.8): α_1 representing the client-to-sever delay, and α_2 representing the server-to-client delay.
5. A Virtual Instrument (VI) was built in LabVIEW to measure the delays α_1 and α_2. The VI runs on a dedicated computer and communicates with a high-precision frequency counter (model PM6669 from Fluke) that measures pulse-width with a resolution of ± 100 ns. The samples are buffered in the local memory of the counter and then transferred in burst-mode to the VI.
6. The system was started. About 1000 samples of α_1 and α_2 were acquired and logged to a file.
7. The network was loaded with two additional computers programmed to exchange large amounts of data through TCP sockets. Step 6 was repeated for the following extra traffic loads: 6%, 12%, 25%, 50% and 70% of network capacity. Figure 2.9 presents the distribution of α_1 for extra traffic loads of 0% and 50%.

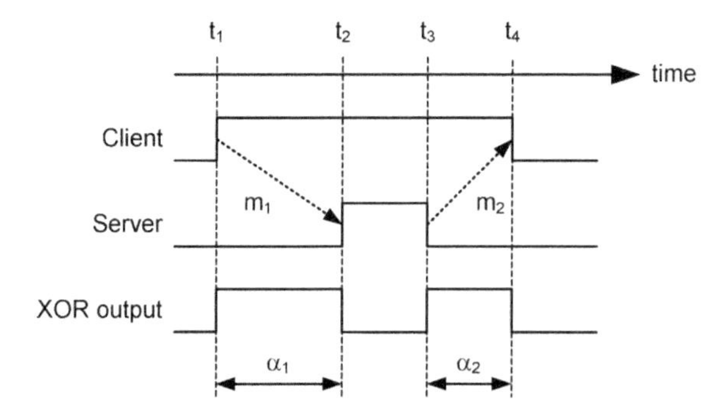

Fig. 2.8 Signals generated to measure client/server delays.

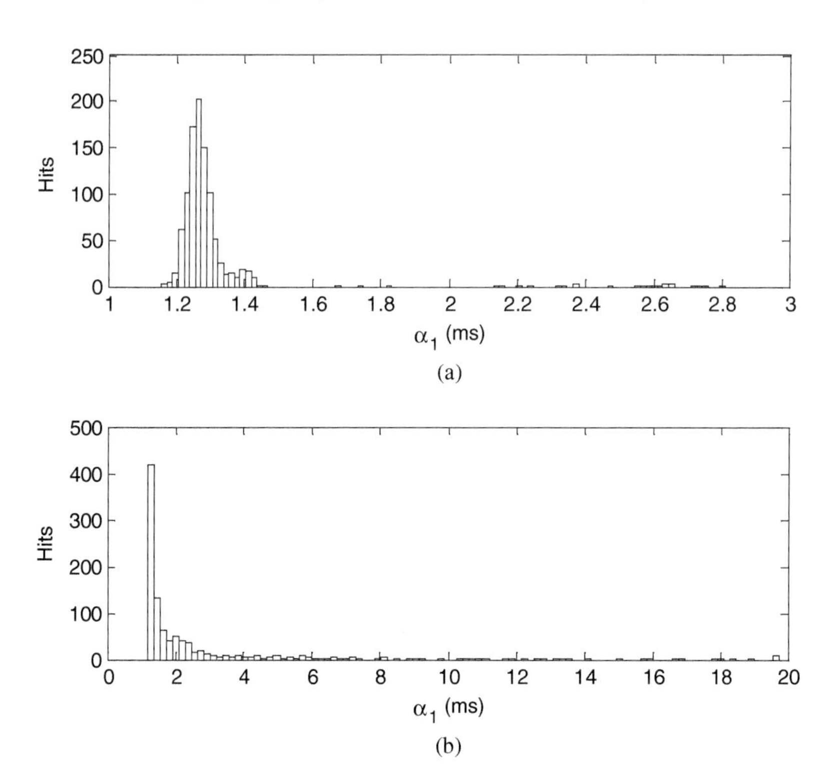

Fig. 2.9 Histogram of α_1 for extra traffic loads of: (a) 0%; and (b) 50%.

Table 2.4. Client/server delays for different traffic loads.

Extra traffic load	α_1 (ms)			α_2 (ms)		
	Mode	Mean	σ	Mode	Mean	σ
0%	1.26	1.30	0.20	0.65	0.62	0.14
6%	1.29	1.46	0.91	0.56	0.81	1.23
12%	1.22	1.81	1.86	0.56	0.97	1.90
25%	1.22	1.81	1.59	0.56	0.97	1.70
50%	1.26	2.63	3.07	0.56	1.37	2.23
70%	1.26	3.06	3.35	0.57	2.04	3.27

Table 2.4 summarizes the results of all experiments. The collected data can be analyzed as follows:

- The client/server delay is in the range of few ms and becomes larger as the contention in the network increases.
- The delay α_1 is larger than the delay α_2 because the client-to-server message is (on average) bigger than the return message.
- For low traffic loads, the delays α_1 and α_2 have Gaussian distributions justified by the random behavior of the Windows scheduler.
- For higher traffic loads, the delays α_1 and α_2 have exponential tails justified by the contention in the network. This exponential behavior is in agreement with the queueing theory [28] used to describe the access to the ethernet medium.

2.5.2.2 Publish/Subscribe Delay

The experiments to evaluate the delay of publish/subscribe communications took place in stations CS2 and ES. The adopted methodology was as follows:

1. The sampling frequency of station CS2 was adjusted to 2 Hz. The control routine of station CS2 was modified to periodically publish an integer variable whose value is incremented by 1 on each loop iteration.
2. A subscriber object was added to the engineering station in order to intercept the published integer. In this scenario, station CS2 is the publisher and station ES is the subscriber.

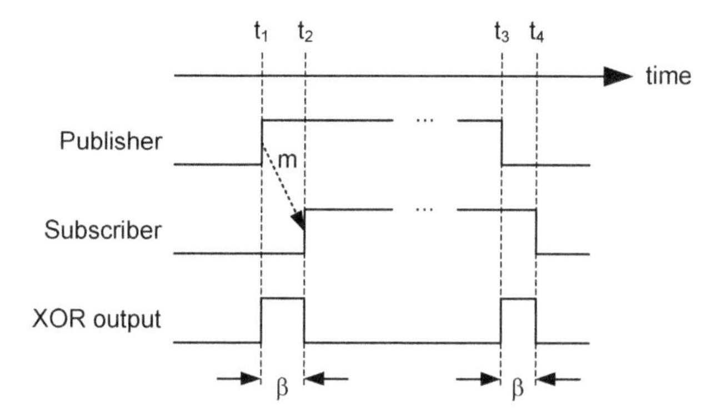

Fig. 2.10 Signals generated to measure the publish/subscribe delay.

3. An instruction was added on the publisher side, just before the remote call, to switch the state of the parallel port data bits. The data bits switch to low/high when the integer variable is even/odd, respectively.
4. The switching process was repeated in the subscriber side: the parallel port data bits switch to low/high when the intercepted integer is even/odd, respectively. This way, a rectangular waveform is created by the subscriber that tracks the one created by the publisher (see Figure 2.10).
5. Both waveforms were connected to a XOR gate to discriminate the delay between them. This delay, represented by β, is the publish/subscribe delay.
6. The system was started. About 1000 samples of β were acquired and logged to a file using the VI mentioned previously.
7. Additional traffic was introduced in the network the same way as described before. Step 6 was repeated for the following extra traffic loads: 6%, 12%, 25%, 50% and 70% of network capacity.

Table 2.5. Publish/subscribe delays for different traffic loads.

Extra traffic load	β(ms)		
	Mode	Mean	σ
0%	1.25	1.25	0.14
6%	1.23	1.31	0.85
12%	1.16	1.43	1.25
25%	1.26	1.46	1.05
50%	1.20	1.76	1.87
70%	2.02	2.18	6.89

Table 2.5 summarizes the results of all experiments. The collected data can be analyzed as follows:

- The publish/subscribe delay is in the range of few ms and becomes longer as the contention in the network increases.
- The distributions of delay β are similar to those verified for delays α_1 and α_2: for low traffic loads, the random behavior of the operating system is noticeable; for higher traffic loads, the contention in the network becomes dominant.
- Sometimes, for very high traffic loads (50% and 70%), the intercepted integer is incremented by 2, meaning that a publication has been lost along the way. This behavior is very rare and is justified by the connectionless nature of UDP.

2.5.3 Behavior of Control Loops

All control loops were extensively tested over their dynamic ranges. For each controller, the setpoint, the process variable and the output were remotely monitored from station ES.

2.5.3.1 Pressure and level loops

The experiments to evaluate the behavior of pressure and level loops took place in station CS1. The adopted methodology was as follows:

1. The sampling frequency of station CS1 was adjusted to 2 Hz.
2. The PI controllers were tuned in advance using the Ziegler–Nichols method based on the open loop step response [29]. The pressure

controller was configured with proportional gain (Kp) = 200 and integral time (Ti) = 6 s. The level controller was configured with Kp = 8 and Ti = 4 s.

3. The system was started with pressure and level setpoints of 2 bar and 50%, respectively. Time was given for all variables to stabilize after which they started to be logged to a file.
4. At t = 60 s, the pressure setpoint was changed to 3 bar.
5. At t = 180 s, the pressure setpoint was changed to 1 bar.
6. At t = 300 s, the pressure setpoint returned to 2 bar.
7. At t = 420 s, the level setpoint was changed to 75%.
8. At t = 540 s, the level setpoint was changed to 20%.
9. At t = 660 s, the level setpoint returned to 50%.
10. At t = 780 s, the logging process was stopped.

The collected data can be analyzed as follows:

- With respect to the pressure loop (see Figure 2.11):
  - At t = 60 s and t = 300 s, the controller opens valve PCV1 (and closes valve PCV2) to increase the pressure inside the

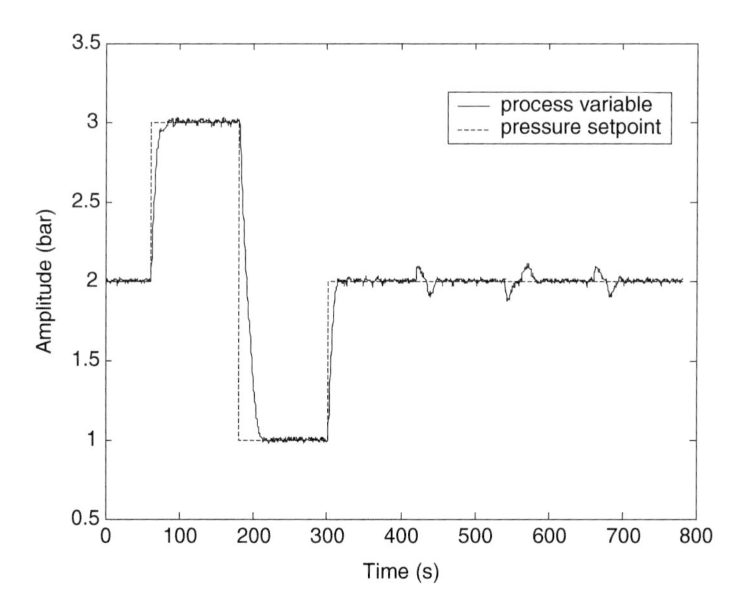

Fig. 2.11 Pressure loop response.

tank C2. The process response has no overshoot and stabilizes after 20 s.

- At t = 180 s, the controller closes valve PCV1 (and opens valve PCV2) to decrease the pressure. The process response has no overshoot and stabilizes after 30 s.
- At t = 420 s and t = 660 s, the pressure has a sudden increase (caused by a level rise) that is quickly canceled by the controller.
- At t = 540 s, the pressure has a sudden decrease (caused by a level fall) that is quickly canceled by the controller.
- The pressure controller is characterized by good tracking capability, no overshoot, short settling time and high immunity to external disturbances (in particular those related with level variations).

• With respect to the level loop (see Figure 2.12):

- At t = 60 s and t = 300 s, the water level inside the tank C2 has a sudden decrease (caused by a pressure rise) that is quickly canceled by the controller.

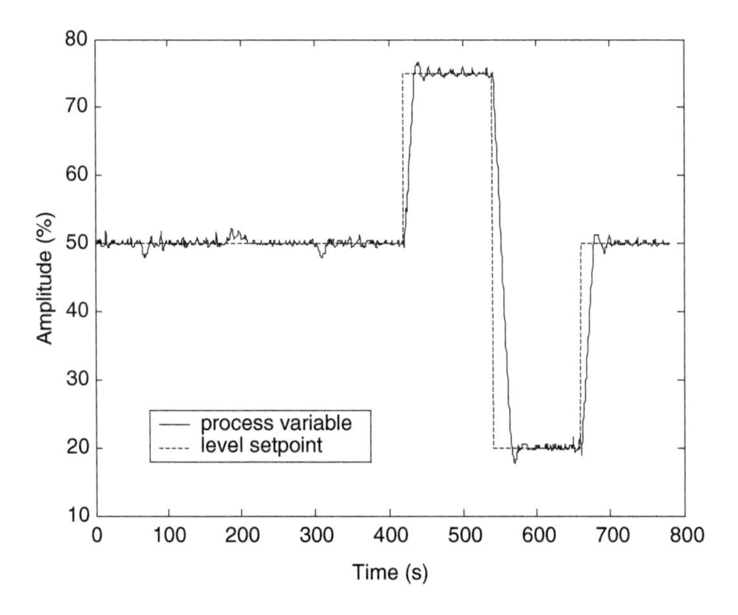

Fig. 2.12 Level loop response.

— At t = 180 s, the level has a sudden increase (caused by a pressure fall) that is quickly canceled by the controller.
— At t = 420 s and t = 660 s, the controller opens valve FCV2 (and closes valve FCV1) to increase the level. The process response has small overshoot and stabilizes after 20 s.
— At t = 540 s, the controller closes valve FCV2 (and opens valve FCV1) to decrease the level. The process response has small overshoot and stabilizes after 30 s.
— The level controller is characterized by good tracking capability, small overshoot, short settling time and high immunity to external disturbances (in particular those related with pressure variations).

2.5.3.2 Temperature loop

The experiments to evaluate the behavior of the temperature loop took place in station CS2. The adopted methodology was as follows:

1. The sampling frequency of station CS2 was adjusted to 2 Hz.
2. The PID controller was tuned in advance using the same method as described before. The temperature controller was configured with Kp = 6, Ti = 10 s and derivative time (Td) = 6 s.
3. The system was started with a temperature setpoint of 35°C. Time was given for all variables to stabilize after which they started to be logged to a file.
4. At t = 60 s, the temperature setpoint was changed to 38°C.
5. At t = 360 s, the temperature setpoint was changed to 32°C.
6. At t = 660 s, the temperature setpoint returned to 35°C.
7. At t = 960 s, the logging process was stopped.

The collected data can be analyzed as follows (see Figure 2.13):

- At t = 60 s, the controller closes valve TCV to increase the temperature of the water entering in the tank C1. As the temperature increases, the derivative component gradually allows the entry of cold water to avoid an excessive overshoot later (anticipation capability). Nevertheless, the process response has an overshoot of 33% and a settling time of 80 s.

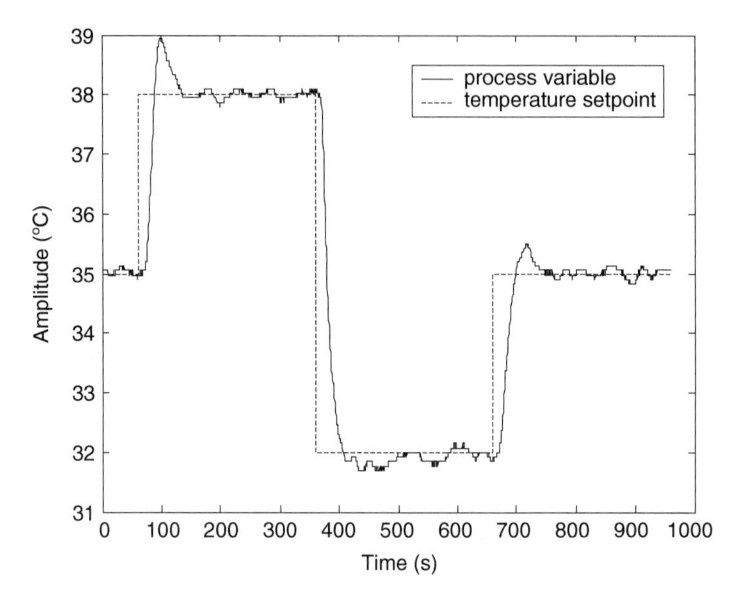

Fig. 2.13 Temperature loop response.

- At t = 360 s, the controller opens valve TCV to decrease the temperature. The process response has smaller overshoot and shorter settling time.
- The instant t = 660 s is similar to the instant t = 60 s, with a slight improvement in terms of overshoot.
- The temperature controller is able to track the setpoint but tends to oscillate due to the long dead time of the process. The anticipation capability introduced by the derivative component is crucial to limit the amplitude of the oscillations.

2.6 Conclusions

The Chapter focus on the SXI platform, a service-oriented middleware solution targeted for instrumentation. The SXI platform was tested in order to know its capabilities and limitations. The tests were conducted on a real plant equipped with all the instrumentation needed to run control loops for pressure, level, flow and temperature, quantities widely found in the process industry. The loops were closed by two control stations and supervised by one engineering station. The three stations were linked together through a 100 Mbit ethernet network. The results obtained led us to the following conclusions:

- The SXI platform is a feasible solution to implement distributed measurement and control systems. The system as a whole, and the control loops in particular, behaved well. All stations worked as expected.
- The SXI platform supports relatively low sampling frequencies (in the range of 10 Hz). This is mainly due to the time it takes to access the transducers. In each access, the hardware of the DAQ board is reconfigured leading to delays that can easily reach tens of milliseconds.
- The SXI platform is not ready for real-time applications. First, because the .NET Framework and the Windows XP operating system are not ready themselves for real-time applications [30, 31]; and second, because communications across ethernet are not deterministic by nature.
- The SXI platform does not guarantee the delivery of publications when the contention in the network is very high.

The limitations of the SXI platform can be overcome by taking the following actions:

1. Implement faster transducer blocks, preferably compliant with the IEEE 1451.0 clause.
2. Quit the .NET Framework and migrate the SXI platform to a real-time environment, such as C++ over the Windows CE operating system [32, 33], or Java over a real-time Java virtual machine [34, 35].
3. Use industrial ethernet switches [36, 37] to make communications deterministic.

The lack of support for real time stands out as the major limitation of the SXI platform at the present, and the first issue to be addressed in the future.

References

[1] David S. Platt, "Introducing Microsoft .NET", 3rd Edition, Microsoft Press, USA-WA, 2003, ISBN 978-0735619180

[2] Jeffrey Ritcher, "Applied Microsoft .NET Framework Programming", Microsoft Press, USA-WA, 2002, ISBN 978-0735614222

[3] Gustavo Alonso, Fabio Casati, Harumi Kuno, Vijay Machiraju, "Web Services — Concepts, Architectures and Applications", Springer, Germany, 2010, ISBN 978-3642078880

[4] Adam Freeman, Allen Jones, "Microsoft .NET XML Web Services Step by Step", Microsoft Press, USA-WA, 2003, ISBN 978-0735617209

[5] `http://msdn2.microsoft.com/en-us/netframework/aa663324.aspx`

[6] Juval Lowy, "Programming WCF Services", O'Reilly, USA, 2007, ISBN 978-0596526993

[7] Justin Smith, "Inside Windows Communication Foundation", Microsoft Press, USA-WA, 2007, ISBN 078-0735623064

[8] `http://ieee1451.nist.gov`

[9] Engene Y. Song, Kang Lee, "Understanding IEEE 1451 — Networked Smart Transducer Interface Standard", IEEE Instrumentation & Measurement Magazine, Vol. 11, No. 2, pp. 11–17, April 2008

[10] "IEEE Std 1451.0 — IEEE Standard for a Smart Transducer Interface for Sensors and Actuators — Common Functions, Communication Protocols, and Transducer Electronic Data Sheet (TEDS) Formats", IEEE Instrumentation and Measurement Society, USA-NY, 2007, ISBN 0-7381-5597-7

[11] "IEEE Std 1451.2 — IEEE Standard for a Smart Transducer Interface for Sensors and Actuators — Transducer to Microprocessor Communication Protocols and Transducer Electronic Data Sheet (TEDS) Formats", IEEE Instrumentation and Measurement Society, USA-NY, 1997, ISBN 1-55937-963-4

[12] "IEEE Std 1451.3 — IEEE Standard for a Smart Transducer Interface for Sensors and Actuators — Digital Communication and Transducer Electronic Data Sheet (TEDS) Formats for Distributed Multidrop Systems", IEEE Instrumentation and Measurement Society, USA-NY, 2004, ISBN 0-7381-3822-3

[13] "IEEE Std 1451.4 — IEEE Standard for a Smart Transducer Interface for Sensors and Actuators — Mixed-Mode Communication Protocols and Transducer Electronic Data Sheet (TEDS) Formats", IEEE Instrumentation and Measurement Society, USA-NY, 2004, ISBN 0-7381-4007-4

[14] "IEEE Std 1451.5 — IEEE Standard for a Smart Transducer Interface for Sensors and Actuators — Wireless Communication Protocols and Transducer Electronic Data Sheet (TEDS) Formats", IEEE Instrumentation and Measurement Society, USA-NY, 2007, ISBN 0-7381-5599-3

[15] "IEEE Std 1451.7 — IEEE Standard for a Smart Transducer Interface for Sensors and Actuators— Transducers to Radio Frequency Identification (RFID) Systems Communication Protocols and Transducer Electronic Data Sheet Formats", IEEE Instrumentation and Measurement Society, USA-NY, 2010, ISBN 978-0-7381-6246-1

[16] "IEEE Std 1451.1 — IEEE Standard for a Smart Transducer Interface for Sensors and Actuators — Network Capable Application Processor (NCAP) Information Model", IEEE Instrumentation and Measurement Society, USA-NY, 1999, ISBN 0-7381-1767-6

[17] Vítor Viegas, J. M. Dias Pereira, P. Silva Girão, "A Brief Tutorial on the IEEE 1451.1 Standard", IEEE Instrumentation & Measurement Magazine, Vol. 11, No. 2, pp. 38–46, April 2008

[18] http://www.w3schools.com/soap/soap_summary.asp
[19] http://msdn.microsoft.com/en-us/library/ms951274.aspx
[20] http://msdn.microsoft.com/en-us/library/ms751494(v=VS.90).aspx
[21] http://www.ni.com/dataacquisition/nidaqmx.htm
[22] http://zone.ni.com/devzone/cda/tut/p/id/4392
[23] http://zone.ni.com/devzone/cda/tut/p/id/2925
[24] http://zone.ni.com/devzone/cda/tut/p/id/4470
[25] http://www.yes01.co.kr/resite/yes01/plint_cat/AT1/thumbs/sect10/te34.htm
[26] http://www.codeproject.com/KB/testing/stopwatch-measure-precise.aspx
[27] http://en.wikipedia.org/wiki/Garbage_collection_(computer_science)
[28] Gerd Keiser, "Chapter 5. Basics of Queueing Theory", "Local Area Networks", pp. 167–201, McGraw-Hill, 1989, ISBN 0071003800
[29] Karl Johan Åström, Richard M. Murray, "Chapter 10. PID Control", "Feedback Systems: An Introduction for Scientists and Engineers", pp. 293–314, Princeton University Press, USA-NJ, 2008, ISBN 978-0691135762
[30] Michael H. Lutz, Phillip A. Laplante, "C# and the .NET Framework: Ready for Real Time?", IEEE Software Magazine, Vol. 20, No. 1, pp. 74–80, February 2003
[31] Martin Timmerman, Jean-Cristophe Monfret, "Windows NT as Real-Time OS?", Real-Time Magazine, Vol. 5, No. 2, pp. 6–13, June 1997
[32] http://www.microsoft.com/windowsembedded/en-us/evaluate/windows-embedded-compact-7.aspx
[33] Douglas Boling, "Programming Windows Embedded CE 6.0, Developer Reference, Fourth Edition", Microsoft Press, USA-WA, 2008, ISBN 978-0735624177
[34] http://java.sun.com/javase/technologies/realtime/index.jsp
[35] Peter C. Dibble, "Real-Time Java Platform Programming, Second Edition", BookSurge Publishing, USA, 2008, ISBN 978-1419656491
[36] http://www.iebmedia.com/ethernet.php
[37] R. A. Hulsebos, "Industrial Ethernet", eBook available in http://www.enodenetworks.com/specials/IndustrialEthernet.epub, 2010.

3

Wireless Sensors Network Testing Environment Based on CAN Protocol*

Gabriel Gîrban and Mircea Popa

Faculty of Automation and Computer Science, Politehnica University of Timisoara

Abstract

Design and development of wireless sensor networks implies software simulations and hardware tests. This chapter depicts the testing environment developed in the lab for WSN nodes energy consumption monitoring and hardware simulation of the nodes behavior in a small scaled network. It is designed to individually control the sensed data and the power supply of the nodes from the network using a wired network of microcontroller based modules that are assigned to one or several sensors. These modules are linked on the same CAN bus with a computer system used to manage both networks (WSN and CAN).

Keywords: WSN, testing, monitoring, testbed

3.1 Introduction

In the last fifteen years, the concepts like smart sensors, smart acquisition systems or intelligent ambient emerge into applications for the large public.

*This work was partially supported by the strategic grant POSDRU/88/1.5/S/50783, Project ID50783 (2009), co-financed by the European Social Fund — Investing in People, within the Sectoral Operational Programme Human Resources Development 2007–2013.

Advanced Distributed Measuring Systems — Exhibits of Application, 35–66.

These types of applications rely on tiny electronic devices that are capable to discover similar systems, being able to collect information about natural or technological systems, primary processing it and wirelessly transmitting the results toward a collecting center. These small devices, called wireless sensors, are based on a sensing module, a wireless transceiver, a low power microcontroller and on a power source. The group of wireless sensors deployed in an area where they can directly or indirectly communicate with each other or with a data collecting center form a wireless network of sensors (WSN).

The wireless sensor networks (WSNs) cover a large spectrum of applications, from environment monitoring to the control of an industrial process, and their size can vary from a wide area wireless sensor network deployed for disasters detection and monitoring to the size of a body area network for health monitoring. These networks are characterized by flexibility, should be less expensive than equivalent wired solutions and the most important, as they are among the cutting edge technologies, a myriad of possible applications can be implemented using them.

The sensing module on a wireless sensor node depends on the application, usually being implemented as an extension board coupled with the core of the wireless node which is also identified as a mote. The purpose of the sensing modules is to detect the variations of specific physical phenomena and their transformation into electrical signals. A mote is composed by the communication module, the data processing module and the power source. The data processing and storage on a mote level is performed by a low power microcontroller. The types of microcontrollers vary from 8 bit to 32 bit data range and from a few kilobytes of internal RAM and ROM memory to several megabytes of internal ROM. There are various ways to supply power on WSN nodes, from different types of batteries to energy harvesting units and capacitors. The communication module is considered to have the most significant impact on energy consumption in a wireless sensor. The medium used for communication depends on the application and there are infrared, optical, ultrasonic or radio communications employed, with the latter as the preferred solution due to its flexibility, although it is the most expensive in terms of energy consumption.

The afore mentioned components of a wireless sensor will be addressed in this chapter when describing the WSN testing environment developed for monitoring the energy consumption and for simulating the sensed data on the network nodes. The principles underlying this testing environment were briefly described in [11].

The next section cast a glance over the existing WSN testing environments, highlighting some of their properties as well as the need for a custom environment that satisfies our requirements. Then, following a short overview, the proposed testing and monitoring system is depicted through the subsystems it contains. Finally, after introducing the application that integrates two subsystems, the limitations of presented prototype and further directions of development are presented within the chapter conclusions.

3.2 WSN Testing Environments

Depending on the design or development phase, there are different types of testing environments that are appropriate. The wireless sensor networks are characterized by a high density of nodes which are usually distributed randomly in certain areas, therefore to design, develop and validate the WSN applications and related communication protocols is quite difficult. It is very important to use hardware devices similar with the application ones and reproduce the real environment even from early stages of the design, in order to obtain accurate data and find the right strategies and network topology. This information can be further used instead of assumptions in calibrating network simulators. Some of the existing solutions for WSN testing are presented below, grouped by the type they belong into testbeds, simulators and emulators.

3.2.1 WSN Testbeds

A testbed is a development platform which allows to perform accurate tests on large software and hardware projects in a replicable manner. WSN testbeds are used to evaluate the software running on the WSN nodes as well as to test their hardware [21], being the preferred solution for experimentation. Several testbed systems are analyzed henceforth.

TWIST (TKN Wireless Indoor Sensor network Testbed) [12] is a flexible testbed architecture developed at the Technische Universität Berlin for indoor wireless sensor network applications. It has 102 TmoteSky nodes and 102 eyesIFX nodes deployed over three floors of a building, with a distance of three meters between the nodes. TWIST allows nodes configuration, network-wide programming, application data collecting, control of the nodes power supply, dynamic selection of flat or hierarchical topologies and controlled injection of node failures.

Motelab [26] is an experimental network deployed at Harvard University to provide public access for development and testing of sensor network applications based on a web interface. The users can reprogram the motes with their own code and all the messages are logged into a database which can be accessed by the user upon job completion in order to be visualized and processed. There are 190 TmoteSky nodes, permanently deployed, each having light, temperature and humidity sensors. The nodes are not running on batteries as they have permanent power supply and are also linked with a central server through an Ethernet connection used for reprogramming and debugging purposes.

MistLab [18] is the MIT testbed deployed indoor. It has 61 nodes (47 Mica-2 nodes and 14 Cricket nodes) spread on the same floor across multiple rooms. It allows reprogramming the motes, scheduling jobs on several nodes or on entire network, and gathering the data logs. Direct communications can be established with individual nodes while running in a job.

Kansei [8] is a complex testbed developed at The Ohio State University. It is focused on sensing and scaling, the nodes being deployed in three different approaches: stationary, portable and mobile. Each node is a pair of XSMs or TmoteSky motes with a Linux based Stargate single board computer. The motes are equipped with several sensors (photocell, PIR, temperature, magnetometer and a microphone) and are connected to the Stargates through a daughter-card interface. The Stargates are connected to a Director PC station using Ethernet interface on stationary arrays or an 802.11 wireless card in case of portable and mobile arrays. There are 210 XSM motes (150 being upgraded with TmoteSky) in the stationary deployment, 50 Trio (TmoteSky) wireless sensors on the portable array and 5 robots as mobile nodes, each having Stargates with both XSM and TmoteSky motes. Kansei provides a web-based scheduling and programming interface, implements tools for sensor data generation (by scaling-up the data received from portable nodes or using parametric and probabilistic sensor models) and allow hybrid simulations using a PC simulation server connected to the stationary or portable arrays.

PowerBench [13] is a 24-node testbed at Delft University of Technology, which is able to gather information about the power consumption of all nodes in parallel. It is using pairs of Linksys NSLU2 devices for each group of 6 nodes. Each node is a TNode(Mica2) — programmer pair with some programmer changes to include an A/D Converter with serial interface, a shunt resistor and on/off switching circuitry. NSLU2 devices are Ethernet backbones with USB2.0 interfaces, the first one being used for programming the motes, controlling their power switch and handling the serial I/O. The shunt resistor is linked in series with the mote and the power supply, the voltage drop variations across this resistor being sampled by the ADC which sends the captured data to the other NSLU2 device. It allows 30 μA (theoretical 16 μA) resolution at a 5 kHz sampling rate and a range of up to 65 mA.

FlockLab [2] is an ETH Zurich testbed architecture which allows the power measurement and stimulation of wireless sensor nodes. Unlike the afore mentioned PowerBench testbed, it is pairing the sensor node with hardware devices capable not only to monitor the nodes behavior but also to stimulate them. The nodes are paired with observer devices using RTC that keep the time over power cycles. These observer devices are capable to perform voltage and current measurements, and may be switched on/off through an external line. A main contribution of this testbed is the time synchronization of the observers with the time base on the server that communicate with all of them and is used for the nodes configuration as well as for collecting and analyzing the test data. There are TinyNodes used as WSN nodes and observer boards based on a Gumstix computer that runs open-embedded Linux.

DSN (Deployment Support Network) [6] can be deployed in the real field as it is based on wireless monitoring devices attached to the WSN nodes. It is developed at ETH Zurich based on BTNode rev.3 to support the development, monitoring, and testing of WSN applications. The wireless monitoring modules are using Bluetooth interface and they are mainly used for WSN nodes event-based monitoring and reprogramming.

3.2.2 WSN Simulators

In early design phase of wireless sensor networks some network simulators are used, especially in finding the proper solutions for communication protocols and network topologies. They are used also for scalability tests when the strategies are verified through simulation of a network with thousand of nodes. In general, the validation of embedded systems requires the use of the real hardware components as the simulations can not cover all the aspects of the environment conditions.

Among the popular WSN simulator are the SensorSim [19], TOSSIM [15], PowerTOSSIM [23] and Castalia [3] simulator based on OMNet++ platform. The SensorSim integrates radio models, battery models, interactions between real nodes and simulated ones and inherits the properties of traditional network simulators (NS-2). TOSSIM is a TinyOS simulator which allows the compilation of the source code and execution on a PC. PowerTOSSIM is actually a package that add power consumption support on the TOSSIM simulator while Castalia provides accurate radio models and integrates a large set of platforms. Also, the EmStar [7] simulation is used within Kansei testbed to calibrate the virtual nodes with real time data and perform hybrid simulations at scale.

3.2.3 WSN Emulators

WSN emulators are usually performing instruction level emulation of the motes microcontrollers. Some of them are able to precisely emulate the radio transceivers, A/D Converters, timers or other peripherals. Such a cycle accurate WSN emulator is Avrora [25] which is scalable but 50% slower than TOSSIM. The AEON [14] is an extension of Avrora emulator that implements a mechanism for accurate prediction of the nodes energy consumption.

3.3 Overview of the CAN Based Testing Environment

The previous section gives a short description of some testing environments that partially emphasize our needs: real time monitoring of the power consumption on the wireless sensors and stimulating the motes with signals similar with the ones produced by the sensing modules in a real environment. The first requirement is fulfilled by the PowerBench while the Flocklab seems to provide support for both of them but the management of observer nodes is less flexible in performing correlations of the stimulation signals with the data sensed by the WSN nodes.

It was not a final goal to design and develop an acquisition and signal generation system but rather to solve the problems of testing the WSN nodes and monitoring their energy consumption in the context described in this section with minimum costs implied. As a reference, the price of a low end data acquisition system with a single channel is more than 1000 EUR (including hardware and software), which is quite prohibitive. Also, if the functionalities, flexibility and extendability of these devices are taken into account, they are not the proper solution for a WSN testing environment. The closest approach matching our constraints in terms of cost and required features is pairing the wireless nodes with monitoring boards which are able to communicate with a central control station. This type of implementation is more flexible allowing additional features like individual motes reprogramming, debugging and strategies verification.

3.3.1 Requirements and Constraints

The WSN testing environment should be able to control the data collected at individual nodes and to monitor the energy consumption at each WSN node that have a pairing equipment with microcontroller. Each of these microcontroller based units assigned to the WSN nodes have to be connected to a wired network used for sending and receiving test related data. An important requirement of the testing system is to be easily extendable for monitoring new sensors nodes.

The charge available in the batteries of the WSN nodes is limited and the impact on energy consumption due to each of the three functional components of a node must be carefully analyzed in order to extend the operation time of that node. Therefore, the equipments that are pairing the WSN nodes must be able to measure the current consumption of the assigned nodes and to implement a realistic battery model capable to provide real time information about the batteries capacity. The current draw variations and related battery discharge rates are useful during design and development of a WSN network as it helps to find the impact of the factors influencing the energy consumption and can be used in improving the network's operating time, too.

Table 3.1. MICAz motes and MTS boards current draw.

Device	Current draw per module/mode		
	Module	Current draw	Mode
Mote	Processing	8 mA	active
Mote	Processing	< 15 μA	sleep
Mote	Communication	19.7 mA	reception
Mote	Communication	17.4 mA	transmission
Mote	Communication	20 μA	idle
Mote	Communication	1 μA	sleep
Sensing	MTS400/MTS420	< 1 mA	accelerometer
Sensing	MTS420	60 mA	GPS

The system characteristics for energy monitoring must satisfy the values shown in Table 3.1 where the technical data of a MICAz node [17] together with some of the compatible MTS extension boards [4] are taken as reference.

A large range of current consumption variations can be observed, from a few microamperes in sleep mode, up to tens of milliamperes in the active mode when using "greedy" sensors. Also, during operation in the active state, different consumption profiles can be distinguished according to the changes of the transceiver state, the sampling rate, heating of electrodes, etc.

To find an optimum between applying energy management of the sensor module, obtaining a minimum flow of the radio messages and an efficient switching rate of the microcontroller in low power modes, it is necessary to reproduce in laboratory the real environment conditions in which the wireless sensors will be used. Given that many of the boundary conditions can not be reproduced in the laboratory, or the costs involved are considerable, the solution is to reproduce the signals generated by sensors under these conditions using a test environment equipped with a signal generator. Information collected by sensors mounted on motes' extension boards is commonly converted into variations of potential difference and detected by A/D Converters. There are also sensors providing information using digital signals, through pulse width modulation or using serial communication lines. Thus, the test environment should be able to replace the sensor during the tests and simulate environmental conditions, otherwise very difficult to be reproduced in the laboratory. The architecture of the testing system should be extensible in order to allow simulation of new types of sensors only through configurations or small software updates.

3.3.2 Testing System Architecture

The primary constraints for the needed WSN testing system are the current draw monitoring, signal generation and possibility to easily extend it to stimulate/monitor new sensors in the network, all with small costs involved. In this context, the system was designed using a CAN serial bus [22] as a communication medium through which monitoring and control boards with a CAN enabled microcontroller are interconnected.[1] There is a computer system connected to the CAN bus using an USB — CAN adapter which is able to send commands to the monitoring boards and to receive the related answers from them. The microcontrollers on monitoring boards are programmed to detect messages on the CAN bus and if these messages are addressed to them, to decode, execute the requested operations and to submit any responses through the same medium of communication. A synthesis of the system architecture is provided in Figure 3.1 while the picture of the prototype is shown in Figure 3.2, where labels were associated to each element.

Figure 3.1 presents a functional point of view of the system as two networks of paired elements, a wired network and a wireless one, each with its own management system. The constructive solution chosen for the wired network allows to connect multiple computer systems, thus the prototype system represented in Figure 3.2 is using two computers connected to the same CAN bus, one of them acting as a common management system for both networks.

The monitoring boards labeled “A” in Figure 3.2 are based on development boards using microcontrollers with a CAN interface[2] and include some additional bipolar transistors, instrumental amplifiers and other discrete components. The CAN bus is labeled with “B” and is connecting the two monitoring boards with the PC systems “F” and “G” through the USB — CAN adapters[3] marked with “C”. The “G” computer station is running a commercial application under Windows XP operating systems and was temporary used as a second device for monitoring the CAN data flow during the prototype validation. The portable system “F” is running a Linux distribution and is used to control the data flow in both networks. There is a single software application

[1] The principles underlying the prototype system described in this chapter are generic, being applicable to any system based on a microcontroller with CAN interface.

[2] ZK-S12-B development boards from SofTek Microsystems with Freescale microcontrollers from MC9S12 family are used in the prototyped system.

[3] The prototype is using USB2CAN adapters from SYS TEC Electronic.

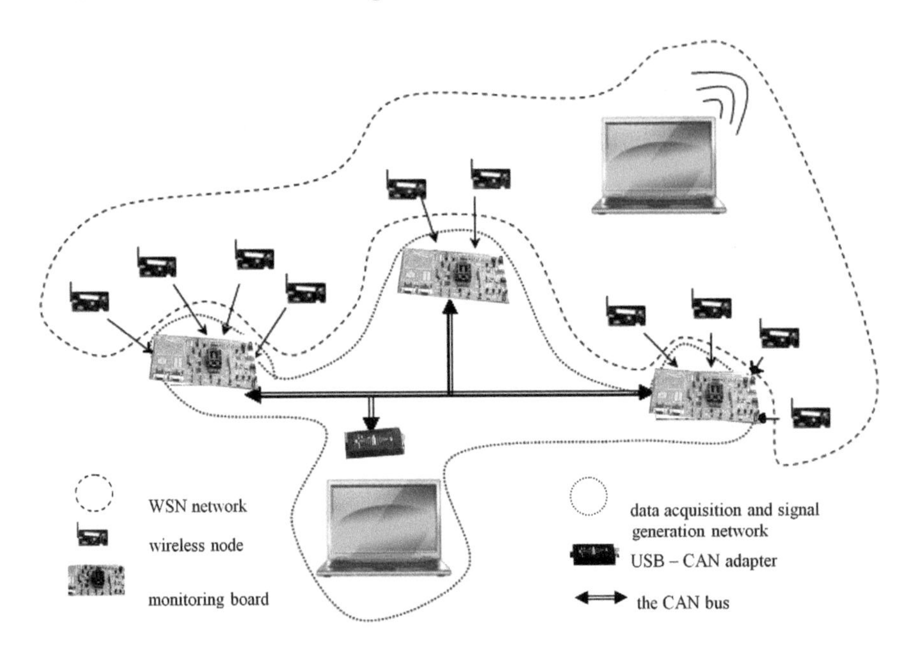

Fig. 3.1 Architecture of the WSN testing environment based on a CAN bus.

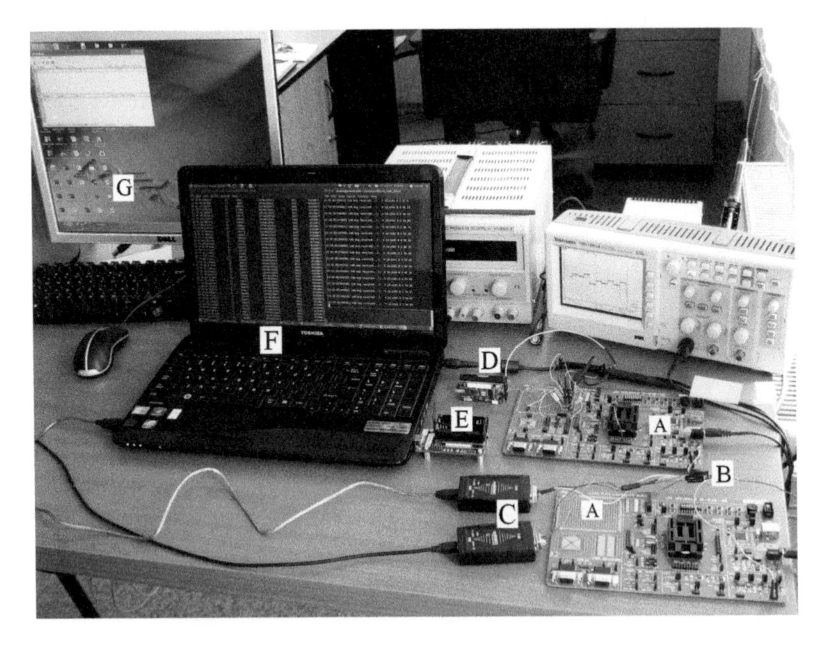

Fig. 3.2 The prototype of the WSN testing environment.

on “F” system which sends commands in the two networks and receives all the messages sent by WSN nodes and by monitoring boards. This portable computer interacts with the nodes from the WSN network, labeled “D”, through the USB gateway interface[4] marked with the label “E”.

3.4 The CAN Network Subsystem

The functionalities of the WSN testing environment are implemented through the elements that form the CAN Network. Among them are the energy consumption monitoring of WSN nodes, the stimulation of the WSN motes with signals simulating the sensed data and management of the tests related data flow.

3.4.1 Energy Consumption Monitoring

The WSN testing environment was designed as a generic data acquisition system, being able to monitor lines with analog signals, digital signals, and even to measure the duration of pulses. These types of signals monitoring can be performed by every monitoring board continuously or sporadically, in response to commands received over the CAN bus. The prototype was build using a Freescale microcontroller [9] and each monitoring board has 8 analog input channels with 10 bit resolution, 8 general purpose channels configured as digital inputs and seven channels reserved for measurement of periodic signals.

The energy consumption monitoring is implemented in a configurable way using commands that are sent over the CAN bus. It is based on measurement of voltage variations across a shunt resistor which is dynamically selected from a set of resistors with different values based on the actual current draw. These resistors are connected between a WSN mote and its power source through bipolar transistors acting as switches. The resistors values are chosen taking into account different consumption profiles and they are selected based on the voltage variations allowed on the motes terminals in order to avoid affecting normal operation of the monitored devices. The principle of connecting these components with WSN nodes is shown in Figure 3.3, where the signal amplification and other filters are not taken into account. WSN nodes are connected

[4]MICAz motes and MIB520 gateway interface are used in the prototype described.

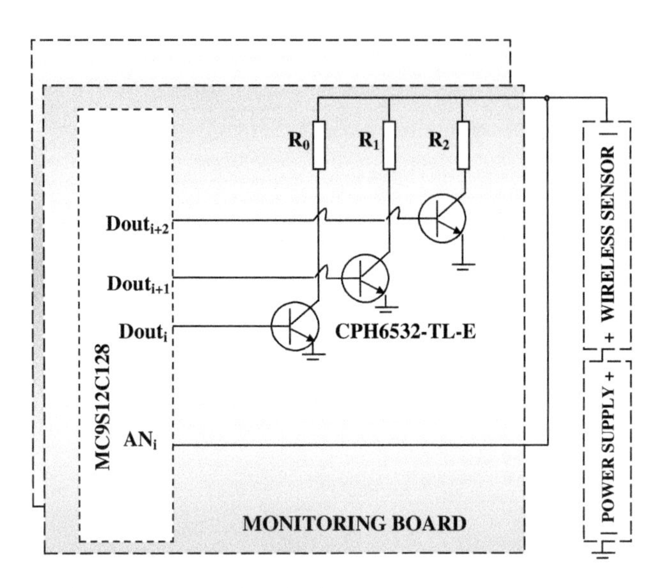

Fig. 3.3 The principle of WSN nodes current draw monitoring.

between the positive terminal of the battery and the resistors. All resistors are linked to a mote terminal and each of them has the other terminal linked to the collector of a specific NPN bipolar transistor with the role of a switch to ground. These switches are controlled through microcontroller channels configured as digital outputs, a single transistor being saturated when the mote is operating in a certain current range. Two transistors are in saturation when switching between adjacent ranges of values so that the node will not be disconnected when the power consumption is changed significantly. When none of these switches is ON, the node will be disconnected from power supply, thus a complete discharge of the battery being simulated.

The system is calibrated using messages sent through the CAN bus, considering the maximum allowable voltage variations on the node power supply without affecting its behavior, the voltage offset introduced by emitter-collector junctions and the resistors values. The software running on the monitoring boards is opening these switches based on threshold values previously sent through the CAN messages. This software uses a data structure for storing the conditions to identify the limits on a certain current range and the channels to be activated when switching to another. Thus, the change is performed automatically based on the value of the specified analog channel, the dynamically

Table 3.2. Relational expressions for automatic signal generation.

operation code	Condition	Operation
0	value of [An_Ch] $= An\_V_x$	
1	value of [An_Ch] $< An\_V_x$	if the condition
2	value of [An_Ch] $\leq An\_V_x$	is true
3	value of [An_Ch] $> An\_V_x$	then:
4	value of [An_Ch] $\geq An\_V_x$	$Dig_{out} Dig\_Ch = Dig\_V$
5	value of [An_Ch] $\neq An\_V_x$	

configured thresholds and relational operators associated with these thresholds. The link between these thresholds assigned to an analog input and the related digital outputs are described in Table 3.2. The first column in the referenced table is presenting the values used for operation encoding, the second column contains the relation between the value detected at An_Ch analog input and a certain threshold An_V_x assign to that analog input, while the third column specifies the digital output channel Dig_Ch and the value Dig_V it will take when the relation specified in the second column will be true. There are maximum 16 entries defined for each analog channel but the number of relations can be increased through software updates if needed.

The voltage variations across the shunt resistor are converted into electric current and the actual battery charge is estimated on the monitoring board taking into account the previously submitted parameters like initial battery capacity, rate capacity and recovery effect, as well as the battery model to be used. For the moment only the linear battery model and a simplified version of the modified Kinetic Battery Model [20] are supported.

3.4.2 Simulation of Sensed Data

The monitoring boards have been configured to interact with the pairing WSN nodes. Therefore, in addition to the digital channels configured as outputs, there are several channels configured as PWM. The behavior of these channels is dynamically controlled through the content of messages received via CAN bus. The PWM signals can be integrated to simulate the behavior of sensors that provide information about a physical process or phenomenon using an analog signal. A short view of the way in which the monitoring boards are linked with the WSN motes to simulate the real environment through signals

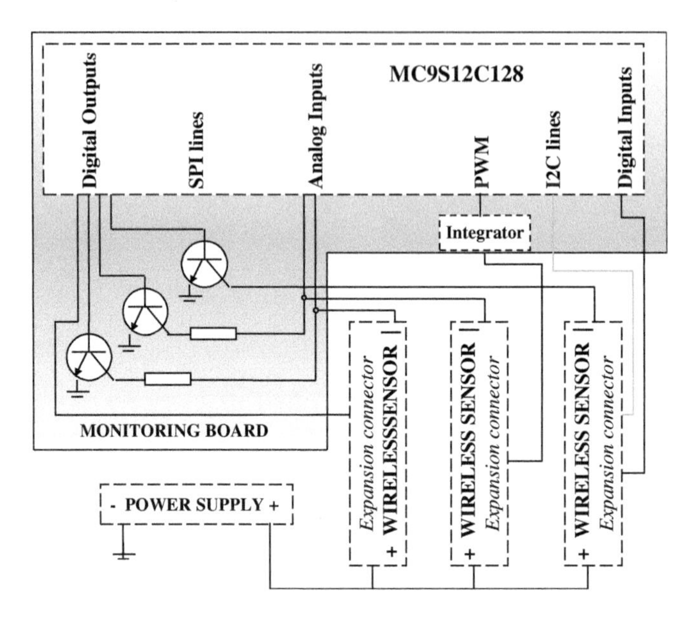

Fig. 3.4 Sensed data simulation using monitoring boards.

generation is given in Figure 3.4. In this figure, the three WSN nodes can be switched on and off through the channels configured as digital outputs. Each WSN node is able to capture information sent over the CAN bus, through the signals generated by the pairing monitoring board. The first node on the left side receives a digital signal, the middle one is stimulated using an analog signal while the sensor on the right receives information through the serial communication lines. There are SPI and I2C interfaces available on the used microcontroller but this feature is not implemented yet on the prototyped system. It can be used in the future for motes reprogramming. The sensor on the right side is also connected to the channels configured as digital inputs, illustrating that it is also possible to monitor the control of the actuators if such devices are connected to the motes.

3.4.3 The CAN Bus

Development of the CAN bus was started by Bosch in 1983 to replace the "point-to-point" communication systems used to connect the electronic control devices in a vehicle. The goal was to create a single serial bus to interconnect all these control devices. The communication protocols were completed in

1986 but the first CAN controller on a chip is available since 1987. CAN version 2.0 [22] used in the WSN testing system described was published in 1991.

Multicasting, broadcasting support and the possibilities to change dynamically the devices in the network without disrupting the operation of other nodes are among the characteristics that led to the choice of this network protocol for the WSN monitoring system. It is not necessary to assign specific addresses to the nodes of a CAN network as there are identifiers assigned to the messages, these being used for messages prioritisation and for messages filtering at destination. The physical access to the CAN bus of the prototyped system is detailed in Figure 3.5. Two wires connected at the ends through 120 ohm resistors are used as communication medium. A CSMA/CA[5] protocol with NRZ[6] coding type is used. The communication lines are constantly monitored by each node in the network, anyone being able to start sending

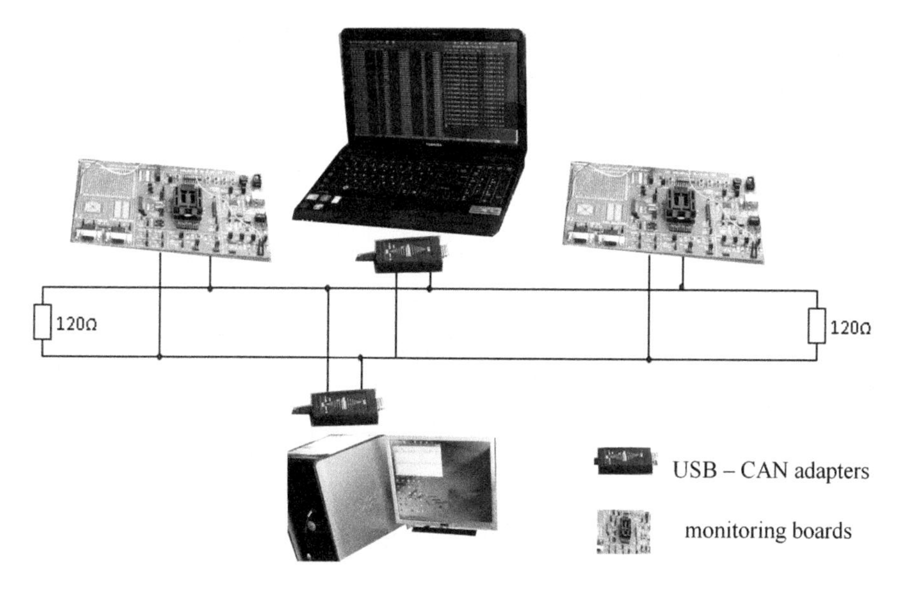

Fig. 3.5 Connecting devices on the CAN network.

[5]Carrier Sense Multiple Access with Collision Avoidance is a medium access protocol in which devices are sending data only when the communication line is free. It is using a collision avoidance scheme for the case in which more devices start transmitting data in the same time.

[6]Non-Return-to-Zero is a binary encoding for digital data transmission where only one voltage level is assigned to each logic value and no resting (zero) state is used.

messages when there is no data flow. If more devices try to send messages at the same time, only one of them will continue sending while the rest of the nodes will detect the collision and will wait for the completion of the pending transmission. There are two levels defined for the medium condition, one dominant and one recessive. To access the physical medium, a wired-AND mechanism is used therefore the recessive level is equivalent with logical value "1" and the dominant level with the logic level "0". The physical environment is in recessive condition only if all nodes transmit the value "1" and collisions arbitration is performed bitwise, in a non-destructive way, a message being accepted by all nodes or by none.

The structure of a CAN 2.0 message is described in Table 3.3. Due to the NRZ coding, synchronization problems may occur for longer sequences when the environment remains in recessive or dominant state (the falling edge is used for synchronization). For resynchronization of the nodes during transmission of a message, for more than 5 consecutive bits with the same polarity, a bit of complementary polarity is automatically inserted. The original message

Table 3.3. The structure of a CAN 2.0 data type message.

Field	Standard ID	Extended ID	Description
SOF	1 bit		start of a message; dominant level
arbitration	11 + 1 bits	11 + 18 + 3 bits	message priority: smaller values ⇒ higher priority; the last RTR bit is dominant; there are 29 bits for extended ID with at least one dominant value in the 7 most significant bits; there are also SRR and IDE recessive bits
control	2 + 4 bits		4 bits for the number of data bytes in message (DLC); for standard ID it includes the IDE dominant bit; r1 and r0 bits are reserved, r1 being the IDE substitute for extended ID
data	0..64 bits		variable number of data bytes as specified by DLC; first byte is the most significant one
CRC	15 + 1 bits		15 bit cyclic redundancy check using the polynomial: $x^{15} + x^{14} + x^{10} + x^8 + x^7 + x^4 + x^3 + 1$; 16th bit is a recessive one indicating the end of the CRC code
ACK	2 bits		ACK and DEL bits. ACK could be dominant on transmission or recessive on reception
EOF	7 bits		and of a data message; there are 7 consecutive recessive bits
IFS	3 bits		3 recessive bits separating two consecutive messages

content is restored on the receiver side, where these synchronization bits are detected and removed. The data transfer rate in a CAN network is configurable by setting the bit time. There are four segments that make up a bit: synchronization, propagation, phase 1 and phase 2, each being a multiple of a time quanta. A breakdown of relations between these segments will be performed when the system configuration is described.

3.4.4 Message Flow on the CAN Bus

The WSN testing environment consists of a central computer system and several monitoring boards, all these elements being connected through a CAN bus. The computer is controlling the monitoring boards behavior and the data flow on the CAN bus by sending requests to certain boards and collecting the data received as response.

The messages exchanged over the CAN bus are configured in extended ID format. Each monitoring board has its own identifier assigned and is able to filter the messages addressed to other equipments. The central system receives all the messages sent over the CAN bus, their filtering being achieved at a higher application level, based on user configurations. The messages sent by the monitoring boards as responses to previous requests have message IDs that include the monitoring board identifier, the type of request and, depending on command, the channel through which the information contained in that message was obtained. Encoding of all these data in the message ID justifies the use of the CAN extended ID message format.

The requests actually implemented on monitoring boards from the prototyped system are summarized in Table 3.4 and are detailed in Appendix using a ($\rightarrow$ *Requests/* $\leftarrow$ *Responses*) formalism. The first data byte in request messages, is always the request code.

In case of a wrong message format or if the parameters are incorrect, an ASCII text will be received from the addressed device, text that will indicate the possible cause of the error.

As presented in the picture of the prototype (Figure 3.2), multiple computer systems can be connected to the CAN bus being able to independently send requests to the monitoring devices but it should be taken into account that all these messages and the related responses will be received by all the other devices from the CAN network.

3.4.5 Prioritisation

According to the CAN protocols, in case of a possible collision, the message with a lower ID will be sent while the devices sending the other messages will try resending them upon completion of the actual transmission. Therefore, the message IDs were assigned or build taking into account two priority layers. The highest priority is assigned to the requests sent by the computer systems using one byte unique IDs assigned to the monitoring boards. In this way, the maximum number of monitoring boards in the system is limited to 256. The second priority layer is differentiating the response messages, based on the type of request, the channel and finally the board ID. Thus, according to the codes in Table 3.4, the highest priority is assigned to the data acquisition on analog channels while the lowest is assigned to the software version responses.

3.4.6 Monitoring Boards Configuration

The application running on monitoring boards is structured in three sections: initialization, interrupt handling and processing loop. The interactions between these sections are presented in Figure 3.6.

In initialization section, the following microcontroller ports are configured:

- 8 analog inputs, configured with a 10 bit resolution, continuous conversion with 8 conversions per sequence;
- 8 general purpose channels configured as digital inputs — PA[0..7];

Table 3.4. Requests implemented on monitoring boards.

Request Code	Description
00h	read of analog input channels
01h	read of digital input channels
02h	measurement of periodic signals
03h	digital output signals generation
04h	PWM signals generation
05h	automatic generation of digital output signals
06h	
07h	WSN nodes consumption monitoring
08h	
09h	analog channel calibration
0Ah	analog data acquisition with sampling time recording
0Bh	time slippage corrections
0C..FEh	reserved for further implementations
FFh	software version on monitoring board

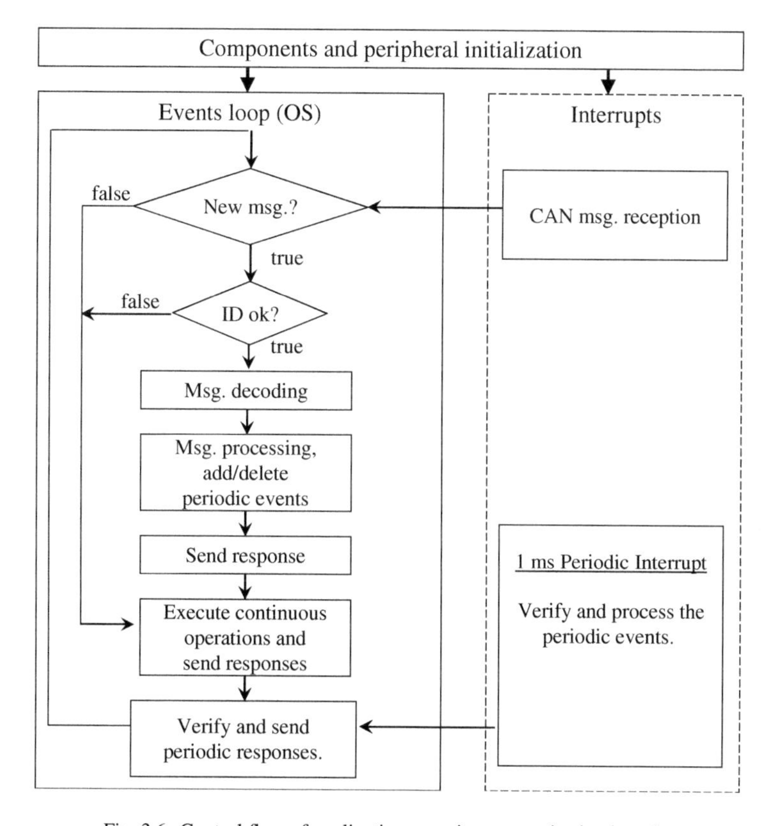

Fig. 3.6 Control flow of applications running on monitoring boards.

- 8 general purpose channels configured as digital outputs — PB[0..7];
- 8 channels configured as PWM[7] — PP[0..7];
- 7 channels assigned for periodic signal measurements — PT[0..6];
- one timer module configured to generate interrupts on each millisecond.

The CAN modules of monitoring boards microcontrollers have been configured to generate interrupts on messages reception. A transfer rate of 500 kbps was chosen and configured following the next steps:

- bit duration is given by $t = 1\,\text{s} \cdot 500000^{-1}\frac{bit}{s} = 2 \cdot 10^{-6}s$;

[7]Pulse-Width Modulation using the PWM8B6CV1 module integrated in some Freescale microcontrollers.

- the following parameters are available on used microcontroller: SJW, BRP, TSEG1 and TSEG2 [9]:

$$t = \frac{BRP_{val}}{f_{CANclk}}(SJW + TSEG1 + TSEG2) \qquad (3.1)$$

- each bit has four time segments and every segment having a certain number of time quanta t_q: synchronization, propagation, phase$_1$ and phase$_2$;
- the bit value is read between the last two segments (shaded area between phase$_1$ and phase$_2$ in Figure 3.7) and duration of the four segments have to comply with the following relations:
  - $8t_q \leq t \leq 25t_q$;
  - $2t_q \leq phase_2 \leq 8t_q$;
  - $4t_q \leq propagation + phase_1 \leq 16t_q$;
  - $SJW \leq 4t_q$ and $SJW \leq minimum\ phase_1, phase_2$;
- the propagation segment acts as a compensation delay due to the signal propagation on the CAN bus. Normal values are between 1 and 14 t_q depending on the cable length. Given that on 500 kbps rate, the length of a line can reach several hundreds of meters and the testing system is deployed in a room, a lower time can be chosen (2 t_q).
- usually, because the ratio between the reference signal and frequency prescale factor is an even number, the number of time units in $phase_1$ and $phase_2$ should satisfy the relations $phase_2 = phase_1$ or $phase_2 = phase_1 + 1$ depending of the parity of the following expression:

$$\frac{t - SJW - propagation}{t_q}. \qquad (3.2)$$

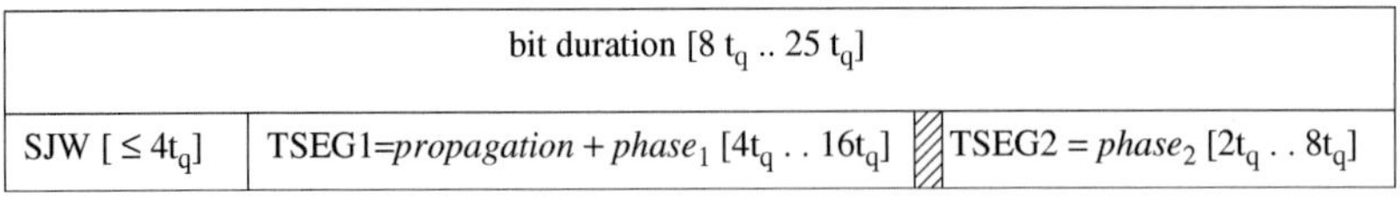

bit duration [8 t_q .. 25 t_q]		
SJW [$\leq 4t_q$]	TSEG1=*propagation* + *phase*$_1$ [4t_q . . 16t_q]	TSEG2 = *phase*$_2$ [2t_q . . 8t_q]

Fig. 3.7 Duration of a CAN bit as a sum of time quanta.

- by substitution in relations 3.1 and 3.2 of $SJW = 1\ t_q$, $propagation = 2\ t_q$, $f_{CANclk} = 16$ MHz, when using a prescale factor $BRP_{val} = 4$, and according to conditions above, the following minimum values are obtained: $phase_1 = 2\ t_q$ and $phase_2 = 3\ t_q$.

3.4.7 CAN Subsystem Improvements

The elements on the presented CAN network act as a test environment with several functions: data acquisition, generation of signals depending on the context and provides multiple sources for data acquisition connected via the same communication medium. The subsystem described in this section was designed as a generic architecture that can be improved in terms of obtained results and performance of the performed measurements, only by replacing the microcontroller used in the prototype with a more efficient one or even with a dedicated system on a chip like the one presented in [16].

3.5 The WSN Network Subsystem

The testing environment was designed to handle wireless sensor networks that can communicate data using an USB gateway. In the prototype presented in this chapter, the wireless network is interfaced with the notebook through the MIB520 adapter [5] with a MICAz mote mounted on top of it (Figure 3.8). The software running on the MICAz mote acts as a relay, receiving the information from the wireless nodes and forwarding it to the computer system or sending to the wireless network the information received from the notebook. This software is based on TinyOS operating system [24] and was written using the nesC [10] programming language. The software architecture is briefly described in Figure 3.9.

The application is divided into four functional modules: initialization, UART interface management, radio interface management and messages processing module. During initialization phase, the two communication modules are activated and the application loop is started. The radio and UART modules will be triggered by the reception of a message on communication interfaces or by requests coming from the processing module when there are messages available to be sent. The processing module, is implemented as a self inserting task into the structure of the system tasks and performs operations like message

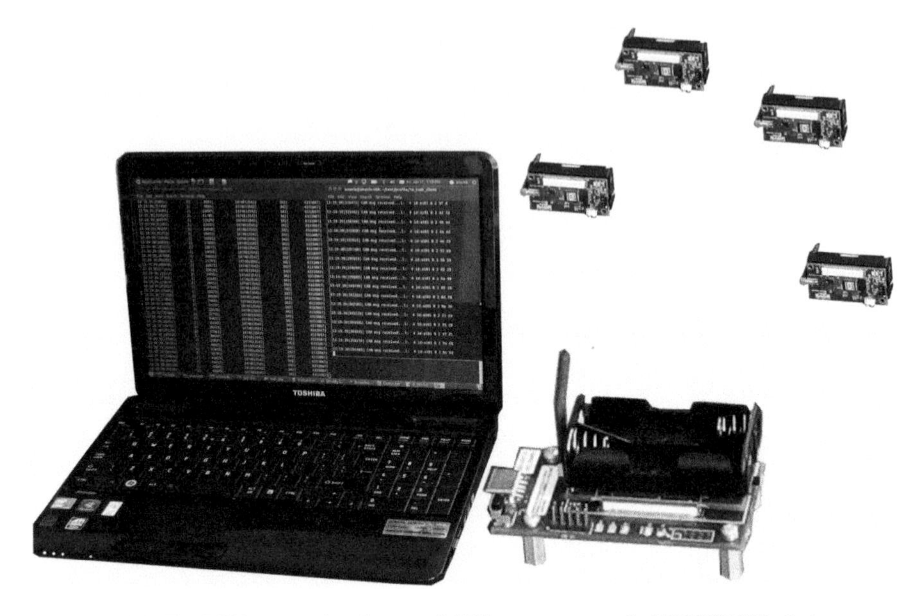

Fig. 3.8 The WSN gateway interface — a MICAz mote on top of a MIB520 USB adapter.

decoding, message processing and formatting as well as data sending on the communication interfaces. All messages received on the radio interface are filtered and if they are addressed to the computer system, are reformatted and send down the serial line. This data flow and the correspondence between the fields of radio messages and the content of the serial messages are described in Figure 3.10. Only the message ID and data fields are unaltered on the WSN gateway interface, the command field being used to control the gateway behavior while the control field of the messages from the radio interface is used to extract (on reception) or encode (on transmission) the message length.

WSN nodes use different types of circuits for clock signal generation, circuits that are usually less precise than those used on monitoring boards. Microcontrollers from the monitoring boards are working on a different frequency than the one used for microcontrollers of the WSN nodes. Therefore, as there are different time references, when detecting the variations in time of physical phenomena, than different information will be obtained about the same phenomenon, in the same time interval, if the sensor is used in parallel with the test system in this purpose. This kind of problems can be exemplified through the integration in time of the WSN node energy consumption. Assuming that the precision of metering circuits is identical on both devices,

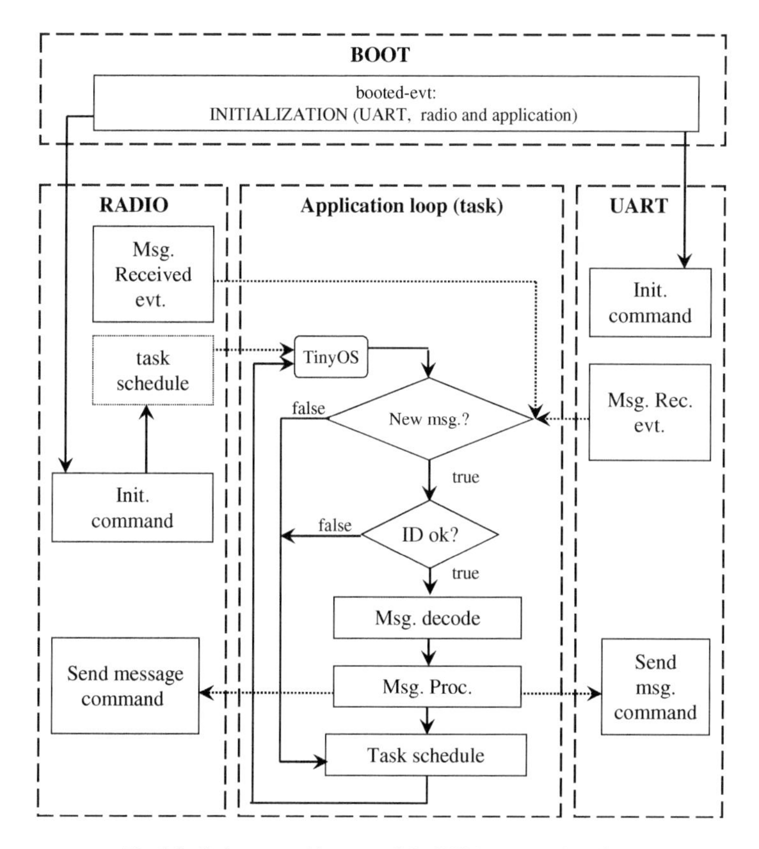

Fig. 3.9 Software architecture of the WSN gateway interface.

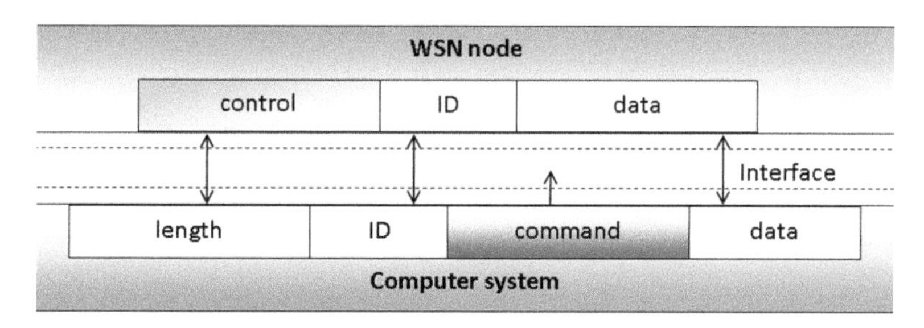

Fig. 3.10 Data flow through the WSN gateway.

when the total consumption is calculated separately for the two systems, an increase of difference between the calculated values will be observed in time. To detect and correct such problems, both WSN sensor and the monitoring system should have accurate time references for the sampled data.

From a functional point of view, the monitoring system and the WSN network are two different networks and in the validation process of the application that is implemented based on the WSN nodes, the motes should execute the application code without being "aware" of the existence of pairing monitoring boards. Thus, a communication through messages exchanged directly between the elements of two networks is not possible and taking into account the slippage that occurs between the clock signals of these elements, a sliding of the time reference is inevitable.

The MC9S12C128C microcontroller used on monitoring boards operate at a frequency of 32 MHz [9] and the external reference is obtained from a crystal in Pierce configuration. MICAz motes are equipped with ATmega128L microcontrollers and for generation of the synchronization signal is able to use an internal oscillator [1]. The nodes frequency can be selectable by software, the default reference being obtained from a 7.3 MHz crystal oscillators.

For determination of the time difference between the MICAz motes and monitoring boards, a node was programmed with a small application that is sending every second a message with the local time, obtained by reading the content of a timer with the resolution of a microsecond. To the monitoring board a 0Ah request (enhanced analog data acquisition) was sent from the computer system, requiring periodic responses every millisecond. The difference obtained is represented in Figure 3.11. The increasing difference followed by

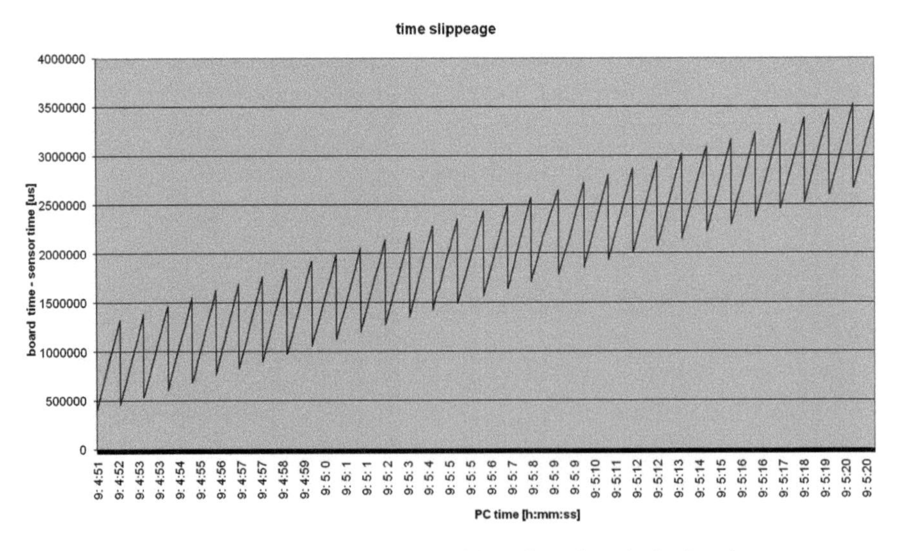

Fig. 3.11 Time slippage between MICAz nodes and monitoring boards.

corrections every second is explained by the low transmission rate of only one message per second on MICAz node, while the time of the monitoring board is updated every millisecond. To reduce the influence of this timing issue, a time correction is used in calculations performed on the MICAz motes. As there are nonlinear dependencies between an accurate time reference and the MICAz time, the correction is based on factors extracted from look-up tables that are filled with values previously obtained in the laboratory.

3.6 Management of the Testing Environment

The previously presented subsystems can be viewed as two independent networks, first using radio communications and the second one based on the CAN communication protocol, whose management is performed through the same software. This section presents the architecture of the networks management software and briefly describe the configurations required for extracting useful information from the data flow between the computer system and the two networks.

The application runs under a Linux distribution and was written in C. An overview of the application structure is given in Figure 3.12.

When the management application is started, it will load the configuration files,[8] set the communication parameters and will create a separate thread for management of the communication with the CAN network. The configuration parameters, their meanings and possible values are described in Table 3.5. All these parameters will be included in the a start-up configuration file in the form of *parameter* = *value* lines.

The main execution block is implemented as a select loop detecting the reception of a message on the two networks or the millisecond event. File descriptor used for reading the WSN messages is obtained by using a system call for accessing the specified USB port while a pipe mechanism is created for the messages received from CAN network. After all events detected on the select loop are processed, the system will continue to execute all the operations previously planned through an execution pipe, based on a FIFO principle. When there are no more operations in the execution pipe, the system is entering again the select loop and will wait for new events.

---

[8] All the configuration files are plain-text files using the ASCII character set.

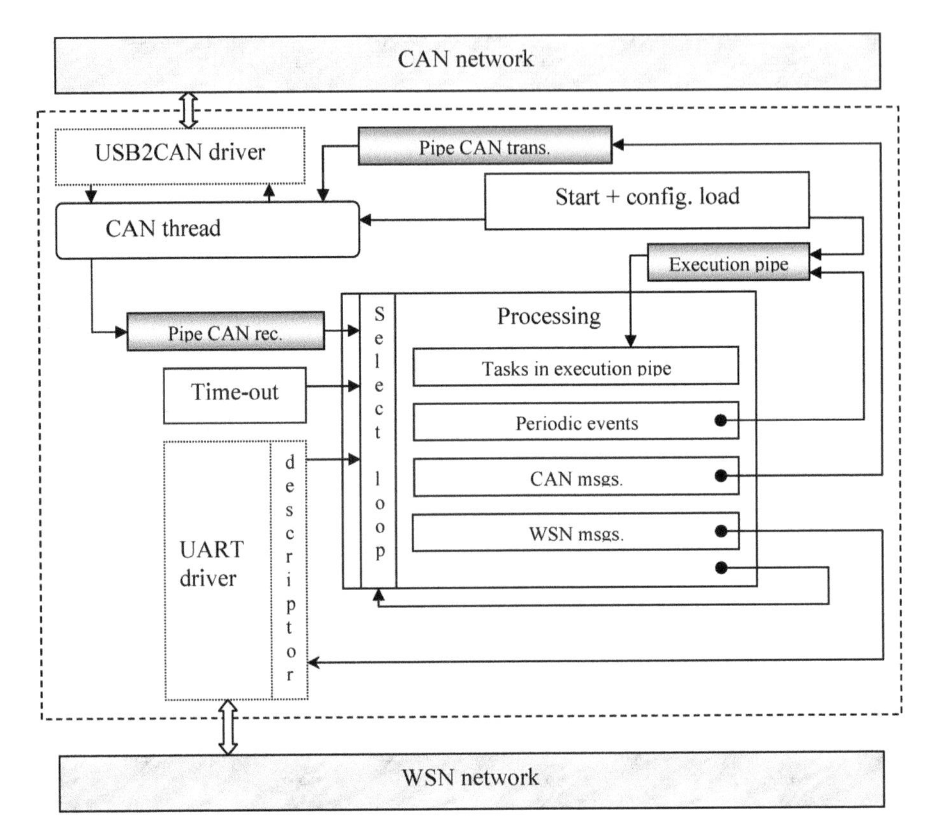

Fig. 3.12 Software architecture for management of the CAN and WSN networks.

Table 3.5. Configuration parameters for management application.

Parameter	Values	Description
CANbaudrate	0 × 0014	1 MBit/s
	0 × 0016	800 kBit/s
	0 × 001C	500 kBit/s
	0 × 011C	250 kBit/s
	0 × 031C	125 kBit/s
CANcmdfile	alphanumeric	name of the file containing messages to be sent on the two networks
LOGcfgfile	alphanumeric	name of the file containing filters for the data to be logged
DebugOn	"Y" / nothing	"Y" is a request for creation of a data flow history
USBbaudrate	numeric	an integer value used for setting the WSN baud rate — implicit value is 57600 bit/s
USBdev	numeric	number of the USB device where the MIB520 adapter is connected

The following operations are performed during events processing:

- event decoding
- for received messages, the information is extracted from the buffer of the event related network;
- for periodic messages, the information is extracted from the buffer of messages ready to be sent;
- the information is processed according to the actual filters;
- the fields of received messages passing the filtering process are sent into a data base;
- if there are messages triggered by periodic events, they are sent to the corresponding network.

Application interacts with the user in the form of configuration files but can be extended with a GUI for transmission and/or reception of data messages.

The software application communicates with the two networks by sending and receiving messages, the main goal being extraction of information about the status of the WSN nodes. Two configuration files are used for communication and data filtering, one containing the messages to be sent and the other one defining the filters for received messages and related fields.

The requests to be sent on the CAN network as well as other messages to be sent to the WSN nodes must specify the targeted network and the type of transmission (periodic or only once at application start-up). All messages are specified as lines in a CSV[9] file, the structure being presented in Table 3.6. Because there are different message structures in the two networks, an internal representation of all messages is defined and described in Figure 3.13.

The fields to be extracted from the received messages are specified in CSV formated file. The file format is shown in Table 3.7 and the result will be a text file with information structured in a tabular form. First two fields are added by default and contain the system time when a message with information from that line was received. The fields following the system time will be those configured in the file, in the order in which they were added. If only several fields are altered by an event, the content of related fields is updated with the

[9]Comma-Separated Values files are used to store data in a tabular, plain-text and human readable form.

Table 3.6. Definition of messages in a CSV file.

Position	Field	Values	Description
1	destination	0	WSN network
		1	CAN network
2	period	numeric	the recurrence of the message in μs
3	ID	numeric	message ID
4	length	numeric	the number of bytes in the message data field
5	data 0		
...		numeric	content of the message data field
5+N	data N		

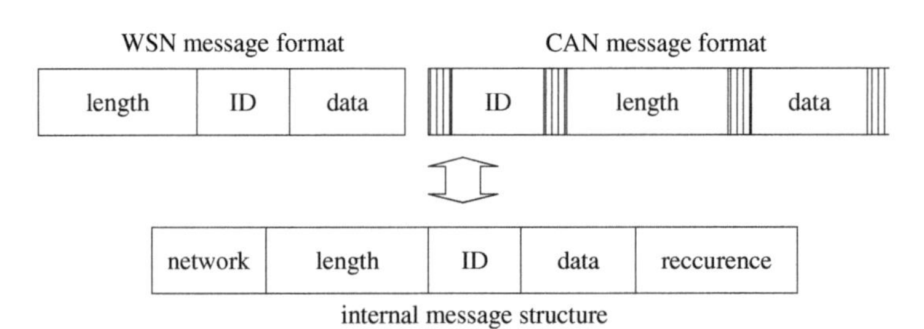

Fig. 3.13 Internal representation of messages.

Table 3.7. Filters configuration in a CSV file.

Position	Field	Values	Description
1	name	alphanumeric	name of the field in the table heading
2	format	alphanumeric	printf's format specifiers
3	data type	numeric	0 specifies a string, otherwise it is the size in bytes of an unsigned integer
4	message source	numeric	0 — WSN network
			1 — CAN network
5	ID	numeric	message ID
6	first byte	numeric	the first byte in the data field to be taken into the account
7	last byte	numeric	the last byte in the data field to be taken into the account

received values, the time of new event will be used and all the other fields will be kept unaltered.

3.7 Conclusions

The system presented in this chapter is a flexible and generic solution used in laboratory for testing and validation of the applications that rely on wireless sensor networks. The system is designed around a CAN network through which messages with commands for the monitoring boards are sent and results of these actions are received. From a functional point of view, there are three types of elements in this system: a computer system that sends and receives messages to/ from a CAN network and a WSN one, WSN nodes and CAN enabled devices.

This testing environment meets the purpose for which it was designed but has some limitations. The currently prototype is using configuration files to send messages and its interactivity is limited by the absence of a graphical interface, even if the CAN messages can be sent using another commercial application with a graphic interface running on other systems simultaneously with the management application.

A limitation of the system developed as a prototype is the lack of facility for nodes reprogramming using the monitoring network, as it is not implemented yet. This can be improved in the future by adding new commands that allow sending CAN messages with the reprogramming content addressed to one or to all monitored WSN nodes.

Another limitation of the system in its current implementation is related to scalability. The maximum transfer rate accepted on the used devices is 1 MBd, therefore the communication is limited at about 8 messages per millisecond. When the number of monitored WSN nodes is increasing, additional monitoring devices will be required on the CAN bus and than the number of messages per time unit, associated to every monitoring board, will decrease as the CAN communication protocol avoids collisions and involves no special arbitration for the access to the medium. To reduce the impact of these shortcomings, the rate transfer on the CAN bus can be improved by using controllers with more efficient CAN adapters, as a 2 MBd transfer rate on the CAN bus allows approximatively 158 messages per millisecond. Other possible directions for extending the system in this case is to create several monitoring networks and to connect the central computer systems via a high speed Ethernet network, but in this case other problems may arise and a top layer application should be introduced to handle the distributed systems.

However, the system was designed to allow future developments, including the possibility to associate a graphical user interface.

Appendix

Requests and Responses Over the CAN Bus

Abbreviations are used instead of numerical values in order to facilitate understanding of their significance.

→ *Dev_ID, SW_ver*

← *Dev_ID + SW_ver, "VER xx.y"*

→ *Dev_ID, Digital_Read, Dig_Ch, Recurrence_in_ms*

← *Dev_ID + Digital_Read + Dig_Ch, Pin_value*

→ *Dev_ID, Digital_Out, Digital_out_Channel, Value*

← *Dev_ID + Digital_Out + Digital_Ch, Pin_value*

→ *Dev_ID, Analog_Read, An_Ch, Recurrence_in_ms*

← *Dev_ID + Analog_Read + An_Ch, High_byte, Low_byte*

→ *Dev_ID, Ext_Analog_Read, An_Ch, Recurrence_in_ms*

← *Dev_ID + Ext_Analog_Read + An_Ch, An_Ch_H, An_Ch_L, Time_H, Time_L*

→ *Dev_ID, PWM_out, Ch, Polarity, Period, DutyCycle*

← *Dev_ID + PWM_out + Ch, Ch*

→ *Dev_ID, CAC_Measure*

→ *Dev_ID CAC_Measure, An_Ch*

→ *Dev_ID CAC_Measure, An_Ch, An_V_H, An_V_L, Op, Dig_Ch, Dig_V*

← *Dev_ID + CAC_Measure, An_Ch*

→ *Dev_ID, Time_norm, An_Ch, Nominator, Denominator*

← *Dev_ID + Time_norm, empty message*

→ *Dev_ID, CC_Prof_Mon, An_Ch, Resistor, OffsetH, OffsetL, Mode, Rec, [Rec]*

← *Dev_ID + CC_Prof_Mon_Mode + An_Ch, Profile_Time*

← *Dev_ID + CC_Prof_Mon_Mode + An_Ch, Profile_Value*

← *Dev_ID + CC_Prof_Mon_Mode + An_Ch, Overall_Consumption*

References

[1] Atmel. 8-bit atmel microcontroller with 128 kbytes in-system programmable flash (rev. 2467x–avr–06/11). *http://www.atmel.com/dyn/resources/prod_documents/doc2467.pdf*, pp. 1–379, 2011.

[2] J. Beutel, R. Lim, A. Meier, L. Thiele, C. Walser, M. Woehrle, and M. Yuecel. Poster abstract: The flocklab testbed architecture. *Proc. 7th ACM Conf. Embedded Networked Sensor Systems (SenSys 2009)*, pp. 415–416, November 2009.

[3] A. Boulis. Castalia: revealing pitfalls in designing distributed algorithms in wsn. In *Proceedings of the 5th international conference on Embedded networked sensor systems*, pp. 407–408, 2007.

[4] Crossbow Technology Inc. Mts–mda series user manual 7430–0020–03 rev. a. *www.xbow.com*, pp. 1–40, April 2004.

[5] Crossbow Technology Inc. Mpr-mib users manual 7430-0021-07 rev. b. *www.xbow.com*, pp. 40–42, June 2006.

[6] M. Dyer, J. Beutel, and L. Meier. Deployment support for wireless sensor networks. *Proc. 4th GI/ITG KuVS Fachgespräch Drahtlose Sensornetze*, pp. 25–28, March 2005.

[7] J. Elson, L. Girod, and D. Estrin. Emstar: Development with high system visibility. *IEEE Wireless Communications*, pp. 70–77, December 2004.

[8] E. Ertin, A. Arora, R. Ramnath, M. Nesterenko, V. Naik, S. Bapat, V. Kulathumani, M. Sridharan, H. Zhang, and H. Cao. Kansei: A testbed for sensing at scale. *Fifth Intl. Conf. on Information Processing in Sensor Networks (IPSN/SPOTS)*, pp. 399–406, April 2006.

[9] Freescale Semiconductor. Mc9s12c family. mc9s12gc family. reference manual rev 01.24. *http://www.freescale.com/filesmicrocontrollers/doc/data_sheet/MC9S12C128V1.pdf*, May 2010.

[10] D. Gay, P. Levis, D. Culler, and E. Brewer. Nesc v1.1 language reference. *http://nescc.sourceforge.net/papers/nesc-ref.pdf*, pp. 1–28, May 2003.

[11] G. Girban and M. Popa. Wsn testing environment with energy consumption monitoring and simulation of sensed data. *Proc. 6th IEEE Intl. Conf. on Intelligent Data Acquisition and Advanced Computing Systems: Technology and Applications (IDAACS 2011)*, pp. 181–185, September 2011.

[12] V. Handziski, A. Kopke, A. Willig, and A. Wolisz. Twist: A scalable and reconfigurable testbed for wireless indoor experiments with sensor networks. *Proc. of the 2nd Intl. Workshop on Multi–hop Ad Hoc Networks: from Theory to Reality, (RealMAN 2006)*, pp. 63–70, May 2006.

[13] I. Haratcherev, G. Halkes, T. Parker, O. Visser, and K. Langendoen. Powerbench: A scalable testbed infrastructure for benchmarking power consumption. *Int. Workshop on Sensor Network Engineering (IWSNE)*, pp. 37–44, June 2008.

[14] O. Landsiedel, K. Wehrle, and S. Gotz. Accurate prediction of power consumption in sensor networks. *IEEE Workshop on Embedded Networked Sensors (EmNetS-II)*, pp. 37–44, May 2011.

[15] P. Levis, N. Lee, M. Welsh, and D. Culler. Tossim: Accurate and scalable simulation of entire tinyos applications. In *Proceedings of SenSys 03, First ACM Conference on Embedded Networked Sensor Systems*, pp. 126–137. ACM, 2003.

[16] MAXIM. 16–bit risc microcontroller–based smart data–acquisition systems. *http://datasheets.maxim-ic.com/en/ds/MAXQ7665-MAXQ7665D.pdf*, pp. 9–3217, 2008.

[17] MEMSIC. Micaz datasheet 6020–0065–05 rev. a. *http://www.memsic.com/products/-wireless-sensor-net/works/wireless-modules.html*, pp. 1–2, 2011.

[18] MIT. Mistlab internet website. *http://mistlab.csail.mit.edu/*, February 2011.

[19] S. Park, A. Savvides, and M.B. Srivastava. Sensorsim: A simulation framework for sensor networks. In *In Proceedings of the 3rd ACM International Workshop on Modeling, Analysis and Simulation of Wireless and Mobile Systems*, pp. 104–111, 2000.

[20] V. Rao, G. Singhal, A. Kumar, and N. Navet. Battery model for embedded systems. In *VLSID'05,18th Intl. Conf. on VLSI Design held jointly with 4th International Conference on Embedded SysSystems*, pp. 105–110, 2007.

[21] O. Rensfelt, F. Hermans, C. Ferm, P. Gunningberg, and L.A. Larzon. An interactive testbed for heterogeneous wireless sensor networks. *Proc. of the 4th Intl Conf. on Distributed Computing in Sensor Systems (DCOSS 2008)*, pp. 529–551, April 2008.

[22] Robert Bosch GmbH. Can specification 2.0. September 1991.

[23] V. Shnayder, M. Hempstead, B. Chen, G. Werner-Allen, and M. Welsh. Simulating the power consumption of large-scale sensor network applications. *Proc. ACM International Conference on Embedded Networked Sensor Systems*, pp. 188–200, 2004.

[24] TinyOS Alliance. Tinyos documentation wiki. *http://docs.tinyos.net/tinywiki/index.php/-Main_Page*, 2011.

[25] B. Titzer, D. Lee, and J. Palsberg. Avrora: Scalable sensor network simulation with precise timing. In *Proc. of the Fourth Int. Conf. on Information Processing in Sensor Networks*, pp. 477–482, 2005.

[26] G. Werner Allen, P. Swieskowski, and M. Welsh. Motelab a wireless sensor network testbed. *In Special Track on Platform Tools and Design Methods for Network Embedded Sensors*, April 2005.

4

Embedded Processing of Acquired Ultrasonic Waveforms for Online Monitoring of Fast Chemical Reactions in Aqueous Solutions

Ahmad S. Afaneh[1] and Alexander Kalashnikov[2]

[1]*Computer Systems Engineering Department, Birzeit University, Birzeit, Palestine, aafaneh@birzeit.edu*
[2]*Department of Electrical and Electronic Engineering, The University of Nottingham, University Park, Nottingham, NG7 2RD, UK alexander.kalashnikov@nottingham.ac.uk*

Abstract

Ultrasonic NDE in the MHz-range is commonly associated with sampling of the waveform of interest with a frequency that makes it difficult to input the samples directly into a waveform processor. Field Programmable Gate Arrays (FPGAs) are frequently used as an intermediate layer to collect the waveform samples before passing them on to the processor for the extraction of the required information. This transfer can take up several seconds, reducing the applicability of the ultrasonic instrument to fast process monitoring. We present the development of an instrument in which high accuracy waveform acquisition logic was coupled with an embedded soft core processor to reduce the above mentioned overhead. This chapter describes the architecture and digital design of the instrument, the results of its verification compared with a commercially available instrument, and the implemented signal processing algorithm. The instrument was applied to monitor an acid — base titration

Advanced Distributed Measuring Systems — Exhibits of Application, 67–93.

reactions with a commercially available pH-meter. Findings showed that it was essential to apply a temperature correction by using an ultrasound wave reflector featuring a water-filled cavity. The instrument achieved five measurements per second of the ultrasound propagation delay in the solution with an estimated sensitivity to chemical changes of approximately 20 ppm by weight.

Keywords: embedded waveform processing; ultrasonic NDE; monitoring of chemical reactions; embedded processor core.

4.1 Introduction

Applications for a wide range of industrial processes and research uses have been found for ultrasonic non-destructive evaluation (NDE) [1]. NDE involves the excitation of an ultrasonic wave in the medium of interest by an ultrasonic transmitter and receiving the propagated wave by an ultrasonic receiver. In the pulse echo mode, the same transducer is used for both transmission and reception. This mode requires placement of a reflector in the medium; the wall of the reaction vessel can act as this reflector. The advantage of the pulse echo mode is the need for a single transducer (which in many cases results in a substantial cost savings) and one-side access to the reaction vessel only.

Ultrasonic NDE has been used successfully for various analytical purposes [2], and these uses can be extended to process monitoring, wherein an ultrasonic instrument constantly measures the ultrasonic properties of a medium in order to check that the process evolves as it should. One of the most important parameters for the monitoring equipment is the update rate, which specifies how frequently the measurements can be taken. Higher update rates are especially desirable for monitoring relatively fast processes. In addition, achieving high measurement accuracy is relevant to the timely detection of subtle changes that may spoil the product or even lead to thermal runaways [3].

MHz-range ultrasound, which is the main focus of this chapter, enables higher temporal and spatial resolution, compared to sub-MHz ultrasound. For this reason, it is widely employed in the analysis of various layered structures and short-range measurement in ultrasonic NDE and medical ultrasound. The received waveforms are digitized by a high-speed analogue-to-digital converter (ADC) operating at 50–200 MHz to achieve several samples per waveform period, which are required for accurate processing. These clock frequencies make it difficult to interface the ADC directly with a waveform processor, such as a microcontroller. Field-Programmable Gate Arrays (FPGAs) are used to collect the required number of samples at high speed for subsequent processing. In most cases, the digitized ultrasonic waveform is passed on to a personal computer (PC) for necessary processing and information extraction. Higher frequency ultrasound (0.1 GHz and above) is predominantly used for acoustic microscopy and research purposes, only because the appropriate transducers and instrumentation are currently rather expensive.

Ultrasound velocity and attenuation in the medium under test are the parameters deduced from the waveforms received by the waveform processor first (Ultrasonic spectroscopy means determining one or both of these parameters across a range of frequencies). These parameters are specific to every material and can be related to its physical properties, such as elastic constants, density, composition, and microstructure in solids, as well as adiabatic compressibility, intermolecular free length, molar sound velocity, molar compressibility, specific acoustic impedance, and molar volume in liquids [2]. Ultrasound velocity is better for characterizing chemical composition at a molecular level than ultrasound attenuation [4]. Ultrasound velocity can be expressed in terms of compressibility of the medium and is extremely sensitive to molecular organization, composition, and intermolecular interactions, which is why ultrasound velocity was found preferable for applications of ultrasonic spectroscopy in the determination of chemical properties of materials, compared to ultrasound attenuation [4].

Both the accuracy and precision (consistency and repeatability) of the ultrasonic NDE measurements depend, in addition to the measurement conditions, on the accuracy of acquired waveforms [5]. Active ultrasonic NDE enables reduction of the additive noise level by exciting the interrogating ultrasonic waves several times and averaging the received signal [6]. The horizontal resolution of the recorded waveforms can generally be improved by increasing the waveform sampling frequency. Equivalent sampling frequency in the case of active NDE can be increased by using several excitations of the waveforms of interest and acquiring the responses with equidistant time shifts within the period of the ADC clock (frame interleaved sampling [7]). Both averaging and frame interleaved sampling were successfully combined in an FPGA based ultrasonic instrument that enabled high accuracy waveform acquisition [8].

Further developments included a custom analogue front end and FPGA board that substantially reduced the cost of the instrument [9].

As an industrial application of ultrasonic NDE, process monitoring is not fully developed to its full potential at the present time. Process monitoring assumes direct or indirect repetitive ultrasonic measurements of the same process parameter (e.g., temperature or density in a process vessel or pipe) for either quality assurance or safety reasons. In addition to particular measurement accuracy and precision figures, process monitoring frequently requires a short response time and short time intervals between successive measurements. Ultrasonic waves travel a few meters per millisecond in liquids and solids and thus can easily meet the first requirement in many cases. The second requirement is compromised in conventional FPGA based ultrasonic NDE instruments by the need to transfer the acquired ultrasonic waveform from the FPGA to a PC for processing. Measurements for the previously developed instrument [8] showed that the transfer of a single waveform from the FPGA built-in memory to a hard drive took about 2 seconds despite the FPGA was directly connected to the internal PCI bus of a PC. This transfer time could be completely eliminated if the FPGA processed the acquired waveform itself, and outputted only a single parameter related to the process behavior, e.g., the ultrasound propagation time. Many researchers reported the applicability of MHz-range ultrasonic NDE to monitoring various chemical and physical mixtures and transformations [10, 11, 12, 13, 14, 15, 16, 17, 18, 19]. One of the biggest obstacles to the applicability of this monitoring is the strong dependence of the ultrasound velocity in water on the temperature [20]. For this reason, ultrasonic measurements are frequently conducted in a rigorously controlled thermal environment (e.g., [10, 11, 13]), but this approach may not be suitable for industrial processes that require heating/cooling stages or for monitoring endothermic and exothermic processes. A better approach would be to monitor the temperature of the monitored substance and then correct the ultrasonic measurements as appropriate [14]. Measurement of the ultrasound propagation delay can be combined with ultrasonic temperature measurements by using a reflector with a water-filled cavity [21].

This chapter presents an embedded active ultrasonic NDE instrument that aimed to considerably increase the update rate of the ultrasonic NDE instrumentation and reduce its cost significantly at the same time by implementing an embedded processing of the acquired waveforms in the ultrasonic instrument itself; and using inexpensive hardware without compromising the high accuracy required by monitoring instruments. The instrument was applied to monitor acid-base titration with immediate temperature corrections, achieving 5 ultrasound propagation delay measurements per second.

Section 4.2 presents the architecture and design features of the developed instrument. Section 4.3 compares waveforms measured by the developed and a commercial instruments to verify the accuracy of the former. Section 4.4 describes the signal processing algorithm implemented by the embedded processor. The experimental setup and procedures are presented in Section 4.5. Section 4.6 discusses the experimental results, and Section 4.7 concludes the chapter.

4.2 Architecture of the High Accuracy Embedded Instrument

FPGA integration with a processor capable of extracting information from ultrasonic waveforms would reduce the board space, number of components, cost, and, ultimately, overall waveform processing time. Two options are available for this integration: hardwired processor cores that are supplied within some FPGAs in addition to the configurable logic resources (e.g., Excalibur devices from Altera, some Virtex-II Pro, Virtex-4, and Virtex-5 series devices from Xilinx, and SmartFusion devices from Actel) and soft-core processors that are configured from the available logic resources of the FPGA (e.g., proprietary processors, such as Nios from Altera, MicroBlaze from Xilinx, and various open-source cores). Although the first approach consumes fewer FPGA resources, if the appropriate processor is readily available, the second approach seems more flexible and future-proof and, therefore, more suitable for research projects. The choice of the FPGA vendor is largely determined by the expertise and development tools available. The Spartan-3E FPGA from Xilinx, available on a Spartan-3E starter board [22], was chosen for this development to satisfy the low cost and fast prototyping requirements.

The hardware of the developed instrument consists of a commercially available, inexpensive, Spartan-3E FPGA board [22] and custom-developed PCB that houses an ADC (AD9215) with the ADC driver (AD8138) and analog front end (AFE)(Figure 4.1) [23]. The AFE includes a transducer's driver (MD1211 and TC6320 for driving transducers with center frequencies up to

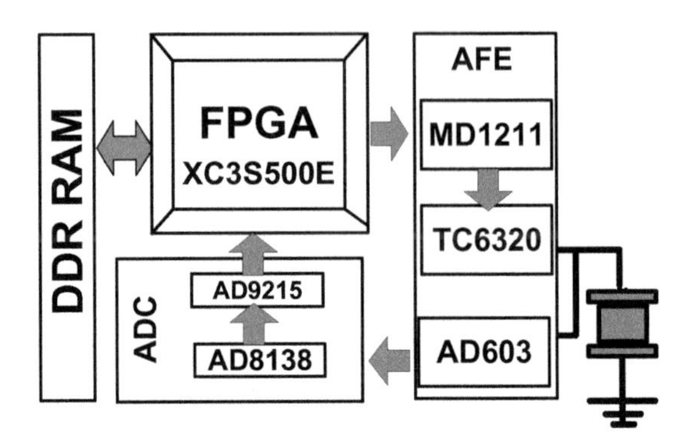

Fig. 4.1 Interaction of the components of the design [23].

20MHz) for excitation of the ultrasonic waves and a variable gain amplifier (AD603) for amplification of the echo. The FPGA triggers the excitation pulse and then collects data from the ADC. The DDR RAM shown in Figure 4.1 is part of the FPGA board, and it is used to store the acquired interleaved waveforms, as discussed later.

An FPGA can implement an entire digital system with substantial processing power by utilizing a soft-core processor with task-specific peripherals (co-processors or intellectual property [IP] cores) for hardware acceleration. In this development, the MicroBlaze processor runs custom-developed software and exchanges data with the co-processors, using shared memory registers. The instrument utilizes both fixed (block random access memory [BRAM] and digital clock managers [DCMs]) and configurable FPGA resources. Figure 4.2 details the constituents of the FPGA digital design and their connections to the external FPGA hardware [23]. The averager, both pulsers, and the DCM controller co-processors were developed using Xilinx's System Generator software tool. Different parts of the digital design were integrated by using the Embedded Development Kit (EDK) from Xilinx. The MicroBlaze processor is responsible for controlling and configuring all the other co-processors in the FPGA. In addition, it controls data transfers between components and processes the acquired waveforms to estimate the ultrasound propagation time. The averager takes data samples from the ADC and provides on-the-fly averaging [6] of the acquired waveforms in the FPGA's BRAM without any MicroBlaze intervention. The required number of averages, N_A, is set by MicroBlaze. DCM controller was the co-processor responsible for controlling the dynamic phase shift interface

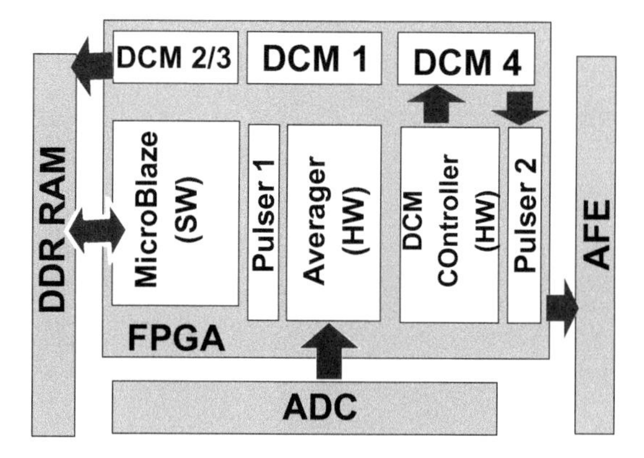

Fig. 4.2 Major components of the FPGA configuration [23].

of the DCM. This controller was the key to achieve interleaved sampling in the design [24]. Development of the DCM controller enabled changing the interleaving factor, I_F, without reconfiguration of the FPGA, which was required for the two-clock interleaving architecture [7]. The I_F is equal to the number of waveforms to be collected and processed by the averager when the excitation signal is delayed by the same amount between the subsequent acquisitions. It is equal to the ratio of the required equivalent sampling frequency f_{eqv} to the sampling frequency of the ADC f_{ADC}:

$$I_F = \frac{f_{eqv}}{f_{ADC}} \tag{4.1}$$

Pulser 1 and pulser 2 generate two independent binary sequences used to maintain relative synchronization between acquiring the echo waveforms and triggering the excitation signal, respectively. The DCM controller delays the pulser 2 excitation sequence by a required amount set by the MicroBlaze. Twenty different delays, equally spread within a single period of the 50-MHz clock signal, were used for clocking the ADC in most experiments, in order to enable $I_F = 20$. The pulser 1 signal triggers the averager to begin the frame acquisition process.

This architecture enabled high accuracy data acquisition by combining averaging with frame interleaved sampling. Operation of the instrument for process monitoring purposes consisted of the following stages [23]:

1. Initialization. MicroBlase sent the required number of averages to Averager and initialised the DCM controller.

2. Averaging. MicroBlaze triggered the excitation of N_A waveforms. Averager averaged these waveforms on-the-fly in the BRAM and notified the MicroBlaze and the DCM controller when the operation was completed.
3. Time shift of the excitation. The averager triggered the DCM controller to set a new time shift, while MicroBlaze copied the averaged waveform samples from the limited capacity FPGA BRAM into the on board capacious DDR RAM, interleaving them with the samples from the other averaged waveforms in the process. It was observed that the DCM finished setting the new time shift before the BRAM-DDR RAM transfer of samples was completed.
4. Acquisition loop. Steps 2 and 3 are repeated I_F times, and the waveform acquisition process is completed after that.
5. Embedded processing. MicroBlaze analyzed the acquired averaged and interleaved waveforms, detected relevant ultrasonic pulses using an adaptive threshold, calculated the ultrasound propagation delays as described in Section 4.4, and communicated these delays to the host. A new monitoring measurement then began from step 2.

The described instrument differs from the previously developed ones ([8, 9]) by using embedded processing, the use of onboard RAM for storing interleaved waveforms, and DCM for delaying excitation, instead of utilizing the two-clock architecture. These features led to the following advantages:

- full compliance of the design with high-level development tools, which makes it more future-proof
- possibility of changing the interleaving factor during operation without the need for FPGA reconfiguration
- compactness of the instrument compared to the previously developed system ([8])
- increase in the size of the data acquisition window to 160 μs compared to the 20 μs achieved in the previous designs ([8, 9])
- cost reduction to approximately \$300 compared to the \$2,000+ ([8]) and \$500+ ([9]) costs of the previously developed instruments, which additionally required a PC for the waveform processing.

4.3 Verification of High Accuracy Acquisition Capabilities of the Developed Instrument

An important step in the development of the instrument was related to the verification of its suitability for high accuracy data acquisition. This was essential because the inclusion of the MicroBlaze to enable embedded processing required re-design of the data acquisition subsystem compared to previously designed instruments [8, 9]. The accuracy of the data acquisition subsystem of the present instrument was tested by applying a train of three pulses (a single pulse period was 1 μs, duty cycle 50%) from pulser 2 (Figure 4.2)) with the repetition frequency of 1 kHz to the input of an RC low pass filter. The output waveform was acquired by using the developed instrument and commercially available PicoScope 5204 with the sampling frequency of 1 GHz and 8 bit resolution [25]. PicoScope acquired the signal in one-shot mode, while the developed instrument with the 10-bit ADC was clocked at 50 MHz and used both averaging ($N_A = 64$) and frame interleaved sampling ($I_F = 20$). As a result, the equivalent sampling frequency for the latter became 1 GHz, and the number of bits in a waveform sample increased to $10 + log_2 N_A = 10 + log_2 64 = 16$ because a single averaged sample was obtained by adding N_A 10 bit samples. The recorded waveform clearly showed better vertical resolution of the developed instrument as it produced smooth waveforms in contrast to the stepped waveforms from the PicoScope (Figure 4.3). However, this increase in clarity increased the measurement time from 1 ms for one waveform required by the PicoScope to 1 ms $\times$ N_A $\times$ $I_F = 1280$ ms. The waveforms captured by the developed ultrasonic NDE instrument did not show rise and fall times as sharp as the ones captured by the PicoScope. This result was due to the limited analogue bandwidth of the developed ultrasonic NDE instrument, which did not require its bandwidth to be on a par with that of a dedicated waveform measurement oscilloscope.

4.4 Embedded Estimation of the Ultrasound Propagation Delays from the Acquired Waveform

The design of the instrument required partitioning of the data acquisition and measurement functions between the hardware and firmware. The former functions were implemented in hardware for speed, and the latter functions were

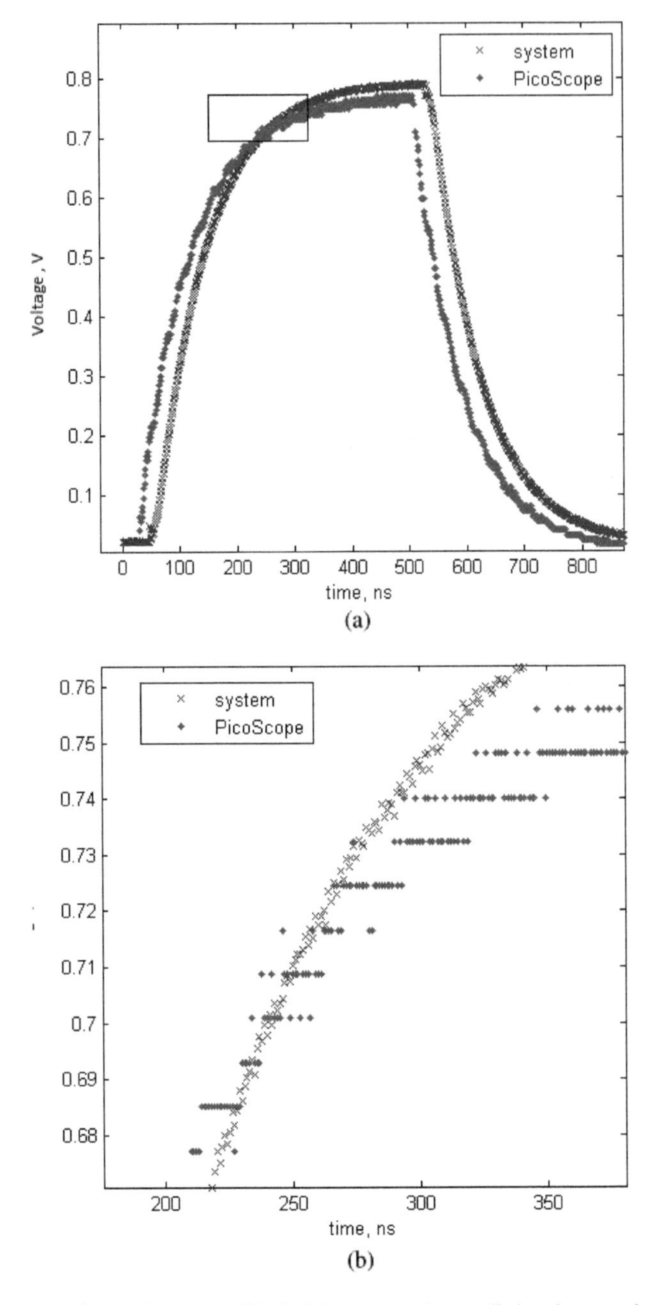

Fig. 4.3 A single pulse captured by both instruments (a-overall view, b-zoomed part).

realized for flexibility as C code for the MicroBlaze, in addition to the code that controlled the hardware. MicroBlaze starts processing when the acquisition of the averaged interleaved waveforms is completed; it estimates the ultrasound propagation delays using an algorithm consisting of four separate stages. In the first stage, the threshold is calculated based on the standard deviation of the additive noise to detect the presence of the echo pulses. During the second stage, the intervals of the waveform that exceed the threshold are found. These intervals are separated into separate pulses in the third stage. Finally, the two samples that bound the first zero crossing are determined for each pulse, and the exact position of this crossing is linearly interpolated. The first zero crossing was selected as the estimate, because it showed the least scatter for a number of ultrasonic waveforms recorded at notionally the same conditions for various solid and liquid media, compared to other zero crossings or their averages. In addition, zero crossings have provided better estimates of the propagation delays estimated for interleaved waveform than center-of-gravity estimates, which worked well for lower sampling frequencies [14]. For the rest of this chapter, the term "zero crossing" will mean first zero crossing. The algorithm was implemented first for the arbitrary locations of the four echo pulses within the complete acquired waveform. However, a substantial period of time was required to process all the received samples, and an application-specific modification of the algorithm was used for the experiments instead. This modification analyzed only the waveforms within the particular time intervals that were determined from the experimental waveforms acquired during preliminary experiments. The comparison showed that this modification did reduce the processing time substantially, without any change in the propagation delay estimates.

4.4.1 Threshold Calculation

Appropriate threshold calculation is essential for the detection of the presence of an echo signal in a particular part of the acquired waveform. Although the threshold value is not used for the delay estimation directly (this can be done by determining the delay at the first crossing of the threshold level by the waveform [26, 27]), the threshold value can still cause a bias in the time of flight measurement. Removing DC offset in the acquired waveform is the first step of its processing. The DC offset is the ZERO level (the middle point of the ADC output) minus the mean value of the signal:

$$\begin{aligned} S_1 &= \sum_i X_i, \quad X \text{ is the acquired waveform} \\ Mean &= \frac{S_1}{N}, \quad N \text{ is the waveform's length} \\ DC_{Offset} &= ZERO - Mean \end{aligned} \tag{4.2}$$

After that the adjusted waveform (WAV) is calculated:

$$WAV = X + DC_{Offset} \tag{4.3}$$

For the rest of this chapter the term "waveform" will refer to the corrected waveform shown in Figure 4.4. Threshold calculation consists of two main steps. The first one is calculating the initial threshold using the three sigma rule:

$$\begin{aligned} \sigma &= \sqrt{\frac{\sum (X_i - Mean)^2}{N}} \\ L &= 3\sigma, \quad L \text{ is the initial threshold} \end{aligned} \tag{4.4}$$

The excitation pulse is excluded from the above calculations to eliminate its effect on the threshold value. Secondly the standard deviation is recalculated

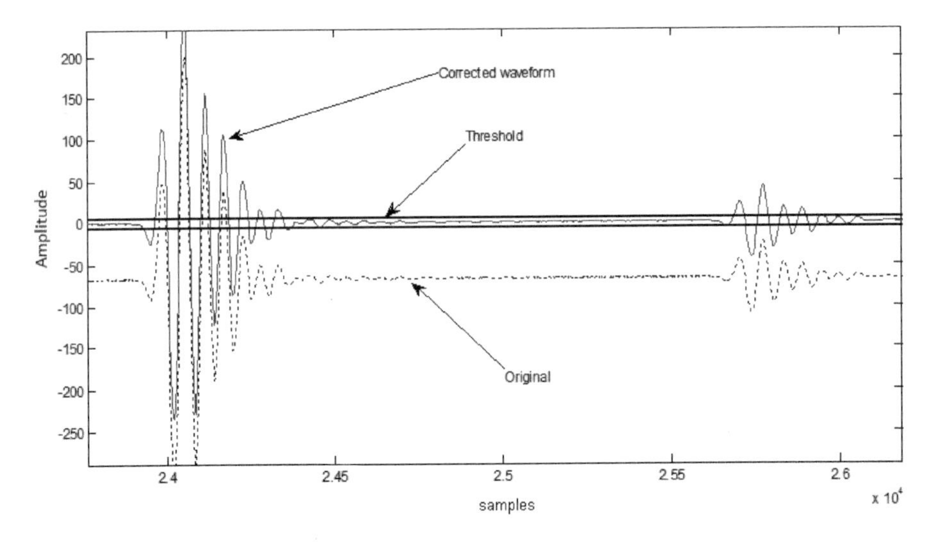

Fig. 4.4 Step1 in the waveform processing algorithm: calculation of the threshold and removing the DC offset.

using only the points between the negative and positive initial thresholds:

$$\sigma_2 = \sqrt{\frac{\sum(X2_i - Mean)^2}{N}}; \quad X2 \subset WAV, \ -L \leq X2(i) \leq L$$
$$\tau = k\sigma_2 \tag{4.5}$$

This step is applied in order to reduce the effect caused by the high peaks of the ultrasound echoes on the initial threshold calculation (Equation (4.4)). The final threshold (τ) is an integer multiple (k) of second standard deviation (σ_2). The value of k was varied depending on the additive noise level compared to the peaks of the echo within the set $k \in \{1, 2\}$. These values led to threshold levels of −22−29 dB below the peak of the pulse.

4.4.2 Detection of the Threshold Crossing Points

Detecting threshold crossing points is the main step in detecting separate ultrasound echoes in the waveform without making the algorithm application dependent. Threshold crossing points are detected by comparing the waveform samples with the negative and positive threshold levels. The output of this step is the list of points where the waveform crosses these levels as shown in Figure 4.5. The detection of these crossing points enables detection of different echo pulses at the next step of the algorithm without any prior knowledge of their locations, hence making the algorithm independent on the measurement

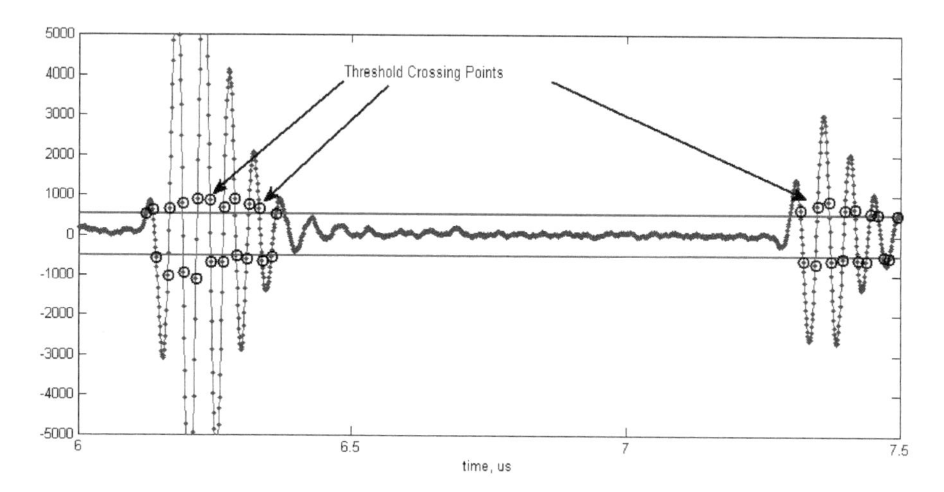

Fig. 4.5 Step 2 in the waveform processing algorithm — detection of the threshold crossing points.

conditions. Detection of the threshold crossing at any point depends on the sign of the previously detected crossing point. For example, if the previous crossing point was detected above the positive threshold, the only acceptable following crossing point should be below the negative threshold. Any subsequent crossings of the positive threshold are ignored by the algorithm as caused by noise.

4.4.3 Detection of the Separate Echoes

The next algorithmic step is the detection of the separate echoes by determining the first threshold crossing point in each echo of interest as shown in Figure 4.6. This is done using a sliding window of four threshold crossing points that moves by one threshold crossing point at a time. A new separate echo is considered detected when a gap of at least G samples is observed between the threshold crossing points:

$$G = 3\left(\frac{1}{\Delta t} \cdot \frac{1}{F_{cf}}\right) \textit{samples} \tag{4.6}$$

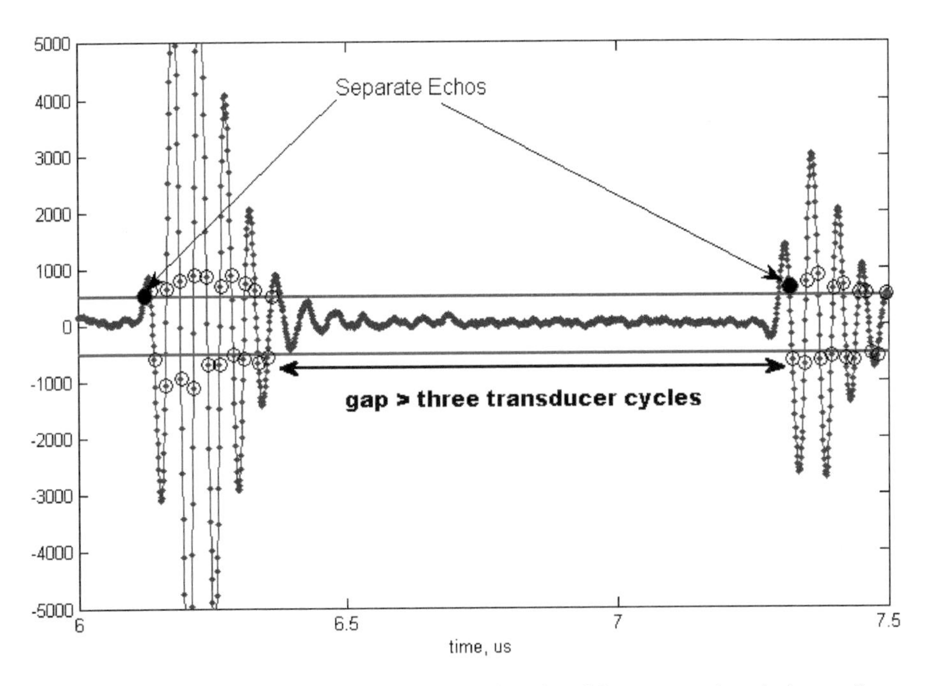

Fig. 4.6 Step 3 of the waveform processing algorithm: detection of the separate echoes in the waveform.

where Δt is the equivalent sampling interval and F_{cf} is the transducer's centre frequency, factor 3 was chosen to set the minimum separation time between different echoes to at least three F_{cf} cycles. Although it was possible to use multiple zero crossing points to estimate F_{cf} out of the experimental records [28], this would lead to unnecessary increases in the processing time. Detection of echoes continues until either all the echoes of interest are detected or the end of the recorded waveform is reached.

4.4.4 Interpolation of Zero Crossing Position

This final step of the processing algorithm takes the array of the threshold crossing points, each representing the beginning of a separate echo, and estimates the position of the first zero crossing for each echo separately. The algorithm analyses all the pairs of the waveform samples starting from the threshold crossing until a pair of samples that enclose a zero crossing is detected ($X_i \times X_{i+1} < 0$) (Figure 4.7a). Estimation of the time location for the first zero crossing point t_z between the two found samples is calculated by linear interpolation for each echo (Figure 4.7b):

$$t_z = \Delta t \left(i + \frac{X_i}{X_i - X_{i+1}} \right) \tag{4.7}$$

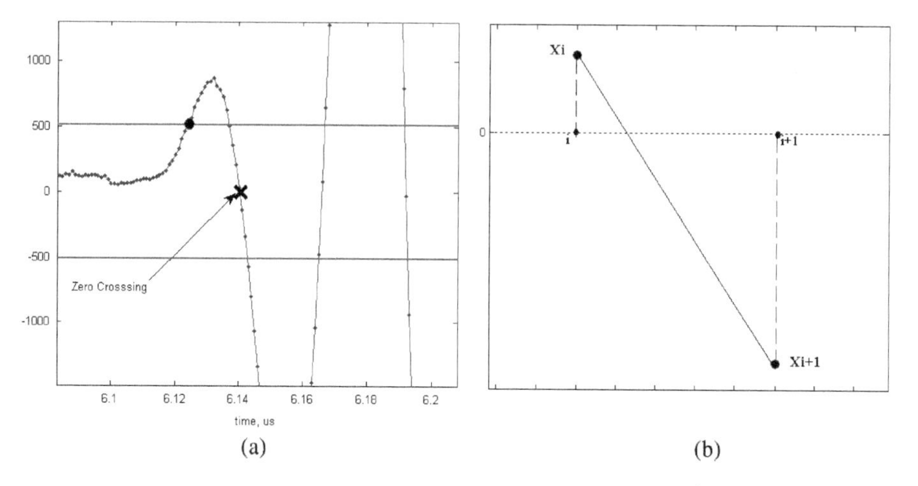

Fig. 4.7 Step4 in the waveform processing algorithm: (a) locating zero Crossing interval; (b) estimating its position between the adjacent samples using linear interpolation.

It was noted that both polynomial and spline interpolations did not provide notable improvements over linear interpolation for an ensemble of the experimental records, but required substantially more time for their calculations.

The fastest measurement rate was achieved when all the pulses of interest (four pulses for the reasons discussed in the next section) were processed in separate relatively narrow windows that reduced the overall number of samples to process in the step 2 above. All the experimental data presented in this chapter were acquired using the excitation pulse repetition rate (PRR) of 10 kHz with $N_A = 64$ and $I_F = 20$. Therefore, the data acquisition time required for a single ultrasound propagation delay measurement was $N_A \times I_F / PRR = 128$ ms. The achieved measurement rate of approximately 5 measurements per second (one measurement in approximately 200 ms) meant that the embedded processing and data communication took about 72 ms, which was considered acceptable.

4.5 Experimental Monitoring of Neutralisation Reactions

To compare the sensitivity of a conventional analytical instrument to the sensitivity of the developed ultrasonic instrument, it was necessary to select a well-understood chemical process that could be conducted safely in an electronic laboratory environment, and that could be conveniently monitored by a well-established analytical method. Acid-base titrations were selected for this purpose; a commercially available pH meter was used as an analytical instrument for their monitoring. Titrations are conventionally used to establish the unknown concentration of an analyte, by adding known amounts of a titrant at the known concentration, until the solution becomes neutral. In this case, the titration was extended far beyond the neutralization point, in order to increase the range for comparison

The experimental setup is shown in Figure 4.8, and the equipment used is listed in Table 4.1. The readings of developed instrument 5 (with probe 4, update rate of 5 measurements per second) and pH meter 3 (with probe 2, update rate of 0.2 measurements per second) were logged by personal computer 7. Ultrasonic probe 4 featured a water-filled cavity that enabled temperature compensation of the measured propagation delays. The titrant was added using automated pipette 1, and the solution was constantly stirred by magnetic stirrer 6.

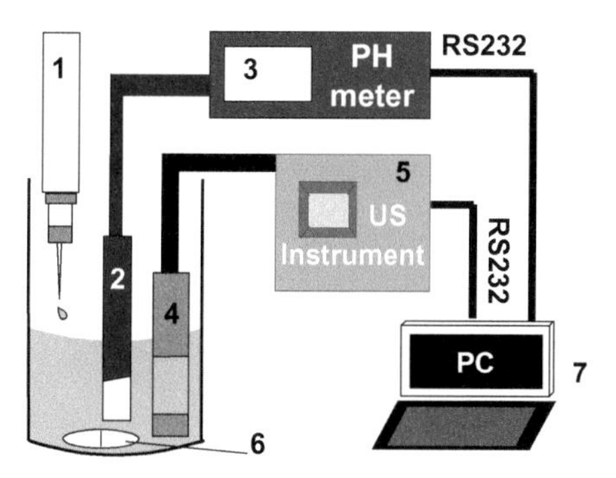

Fig. 4.8 Experimental setup.

Table 4.1. Equipment used for the experiments (Figure 4.8).

No.	Purpose	Model and manufacturer	Comments
1	Liquid dispensing	Multipette Pro (Eppendorf, Germany)	The volume of the titrant droplet was programmed. Releases of drops were triggered manually at set time
2,3	pH and temperature monitoring	340i pH meter (WTW, Germany)	Measurement rate of 1 measurement per 5 seconds. Continuous output to a PC via RS-232 interface.
4	Holding the ultrasonic transducer and sensing temperature ultrasonically	N/A	A dipstick-type probe with a 20 MHz Panametrics transducer V316-SU that featured enclosed isolated cavity filled with deionised water
5	Monitoring of the ultrasound propagation delay	N/A	The custom developed ultrasonic instrument described above (5 measurements of the ultrasound propagation delay per second)
6	Homogenization of the solution	N/A	
7	Data logging	PC	A laboratory PC with one built in COM port, the second port was provided using a USB to RS-232 converter. Custom software was used to save both pH meter and the ultrasonic instrument data into a MATLAB® data files

The experiments were conducted using strong acids, weak acids, strong bases and weak bases. The selected set of reactions (Table 4.2) covered all the possible combinations among strong/weak acids and strong/weak bases, and additionally included further dilution of an aqueous solution. Both pH values and ultrasonic delays were continuously measured during the titrations.

Table 4.2. Summary of monitored reactions; all chemicals had 0.1 M concentration.

Reaction	Analyte	Titrant	Titration rate
$HCl + NaOH \rightarrow NaCl + H_2O$	120 mL H_2O 20 mL NaOH	30 mL HCl	0.5 mL/60s
$HCl + NaOH \rightarrow NaCl + H_2O$	110 mL NaOH	200 mL HCl	1 mL/30 s
$HCl + NaOH \rightarrow NaCl + H_2O$	110 mL NaOH	200 mL HCl	10 mL/300 s
$NaOH + CH_3COOH \rightarrow CH_3COONa + H_2O$	110 mL NaOH	200 mL CH_3COOH	10 mL/300 s
$HCl + NH_3 \rightarrow NH4Cl$	110 mL NH_3	150 mL HCl	10 mL/300 s
$CH_3COOH + NH_3 \rightarrow CH_3COONH_4 + H_2O$	110 mL NH_3	130 mL CH_3COO	10 mL/300 s
$NaOH + H_2O \rightarrow Na^+ + OH^-$	150 ml NaOH	150 mL H_2O	10 ml/3000 s
$CH_3COOH + H_2O \rightleftharpoons CH3COO^- + H3O^+$	150 ml CH_3COOH	150 mL H_2O	10 ml/3000 s
$NH_3 + H_2O \rightleftharpoons NH_4^+ + OH^-$	150 ml NH_3	150 mL H_2O	10 ml/3000 s

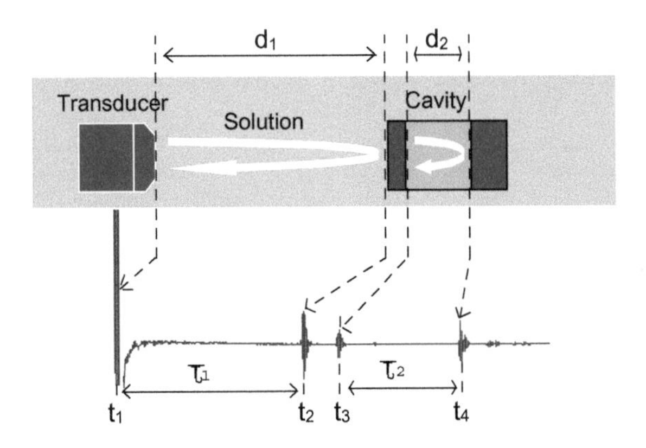

Fig. 4.9 Association of the recorded pulses with the interfaces of the dipstick probe.

The amount of titrant dispensed and the titration rates were varied in some experiments in order to establish their influence on the results.

As mentioned above, the strong dependence of ultrasound velocity on temperature complicates discrimination between changes caused by temperature against changes in chemical composition. Measurement of the ultrasound propagation delay can be combined with ultrasonic temperature measurements by using a reflector with a water-filled cavity [21]; this approach was used for the dipstick probe used in the experiments. The probe contained a 20-MHz transducer and a reflector with a sealed cavity filled with deionized water (Figure 4.9).

The changes to the part of the ultrasound waveform that corresponded to the water cavity were used as an indicator of temperature change in the solution under test. The barrier separating the water from the solution under test had an acoustic impedance value comparable to the acoustic impedance of the water; therefore, most of the ultrasound pulse's energy traveled through toward the other boundary of the cavity. The space between the transducer and the barrier was filled with the solution under test.

The recorded waveforms contained a number of pulses that were reflected from various interfaces of the dipstick probe (Figure 4.9), and only four of them with the delays $t_1, \ldots, t_4$ were used for the ultrasound propagation delay measurements. The excitation pulse (delay t_1 from the time origin) applied to the transducer from the transducer's driver was converted to an ultrasound pulse that travelled distance d1 to the first liquid-solid interface (i.e. the barrier separating the water from the solution under test) and then back in the solution reaching the transducer with the overall delay of t_2. This pulse also partially penetrated the barrier, and was partly reflected from the "reflector-deionised water in the cavity" interface, arriving at the transducer with delay t_3. The final pulse of interest was reflected from the second boundary of the water filled cavity, and arrived at the transducer with delay t_4. The respective differences between these delays gave the experimental values for the ultrasound propagation delays in the solution ($\tau_1 = t_2 - t_1$) and cavity ($\tau_2 = t_4 - t_3$). Equations (4.8) and (4.9) show the relation between the experimental values and the ultrasound velocity during the course of the experiment:

$$\tau_1 = \frac{2d_1}{c} = \frac{2d_1}{c_0 + \Delta c_c - \Delta c_t} \approx \frac{2d_1}{c_0 + \Delta c_c}\left(1 + \frac{\Delta c_t}{c_0 + \Delta c_c}\right) \approx \frac{2d_1}{c_0 + \Delta c_c}\left(1 + \frac{\Delta c_t}{c_0}\right) \tag{4.8}$$

$$\tau_2 = \frac{2d_2}{c} = \frac{2d_2}{c_0 - \Delta c_t} \approx \frac{2d_2}{c_0}\left(1 + \frac{\Delta c_t}{c_0}\right) \tag{4.9}$$

where c is the actual ultrasonic velocity for every measured ultrasound propagation delay. This velocity was different from the ultrasound velocity in deionised water at, say, 25°C c_0 due to the influence of temperature (Δc_t, affected both the solution and the cavity) and chemical composition for the solution only (Δc_c). We assume that temperature changes caused the same deviations of the ultrasound velocity for both the solution and the cavity, because deionised water in the cavity was isolated from the solution by the stainless steel case with high thermal conductivity and small specific heat capacity. Therefore, any temperature changes in the solution (caused by either ambient temperature or energy release/consumption during the titration) affected ultrasound velocity in the cavity quickly.

The above assertion implies that the temperature dependence of the ultrasound velocity in the solution is the same as that in deionized water. Although this assertion is incorrect in the general sense, there is experimental evidence that it holds for low-concentration solutions [29] used in the reported experiments.

4.6 Experimental Results

Ultrasound attenuation is the best variable for characterizing dispersed phase composition and particle size in aqueous solutions; on the other hand, ultrasound velocity is better for characterizing chemical composition at a molecular level [4]. Ultrasonic velocity can be expressed in terms of compressibility. This parameter is extremely sensitive to the molecular organization, composition and intermolecular interactions in the analysed medium. That is why ultrasonic velocity is used as the basis for applications of ultrasonic spectroscopy to the determination of chemical properties of materials. High sensitivity of ultrasound velocity to temperatur is one of the primary factors why ultrasound attenuation is frequently preferred for direct measurement of material properties (e.g. concentration of particles) [2]. It was expected that the use of temperature compensation described previously could extend applicability of ultrasound velocity measurements for monitoring chemical processes.

A typical experimental waveform directly recorded at the ultrasonic transducer is presented in Figure 4.10 The ultrasound propagation delays in the solution and the cavity were calculated from these waveforms by the embedded MicroBlaze processor, as described in Section 4.4, and they are presented in the relevant figures below. One of the objectives of the experiments that were conducted was to determine whether the maximum of the ultrasound propagation delay would occur at the neutralization endpoint (pH=7). Such a maximum had been observed previously, during monitoring of aluminum ion hydrolysis [21]. For the present set of experimental results, this was

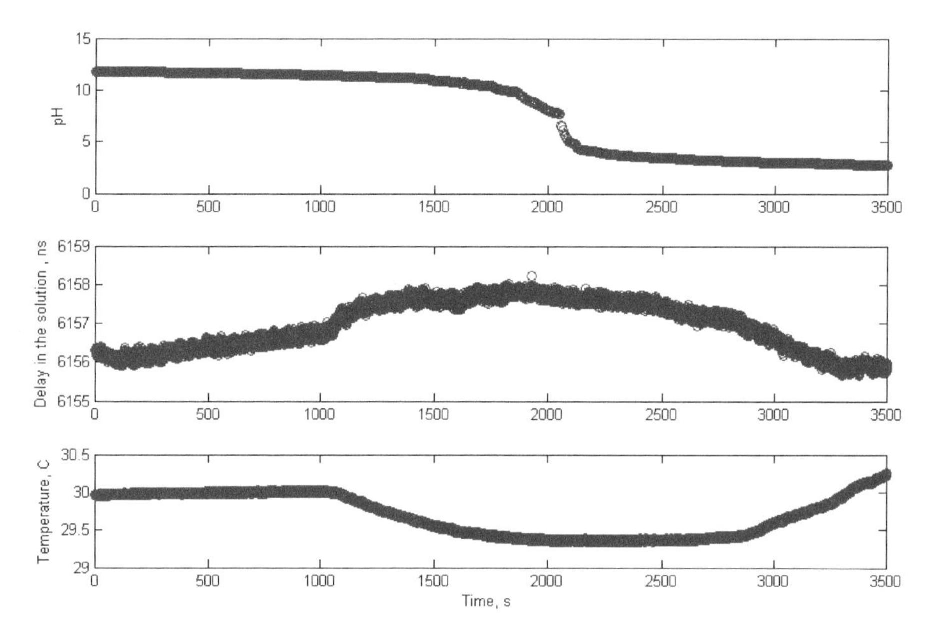

Fig. 4.10 Instrument readings for the case when 1 mL of HCl was added every 60 seconds to a low concentration solution of NaOH, pH data (top), corrected ultrasound delay (middle) and the temperature profile of the experiment (bottom).

not observed for the majority of the collected data. Therefore, the appearance of the above-mentioned maximum was likely related to gel formation, rather than to neutralization of the solution. For this reason, only a subset of the collected experimental data for the strong acid-strong base reaction ($HCl + NaOH \rightarrow NaCl + H_2O$) is discussed in this section. The complete experimental dataset has been presented elsewhere [23].

For the first experiment, a low-concentration (0.02M) solution of NaOH was obtained by diluting a standard 0.1M solution with deionized water. Then, 1 mL of HCl at 0.1 M was added to the solution every 60 seconds. Figure 4.10 shows that the ultrasound propagation delay changed only slightly over the course of the experiment, despite the substantial pH change; the maximum of the former was observed as pH = 7.

A closer look at the experimental data shows that this change was likely caused by the temperature change during the experiment. The other observation is that the changes in ultrasound propagation delay after the application of temperature correction vary within small limits, most likely due to noise

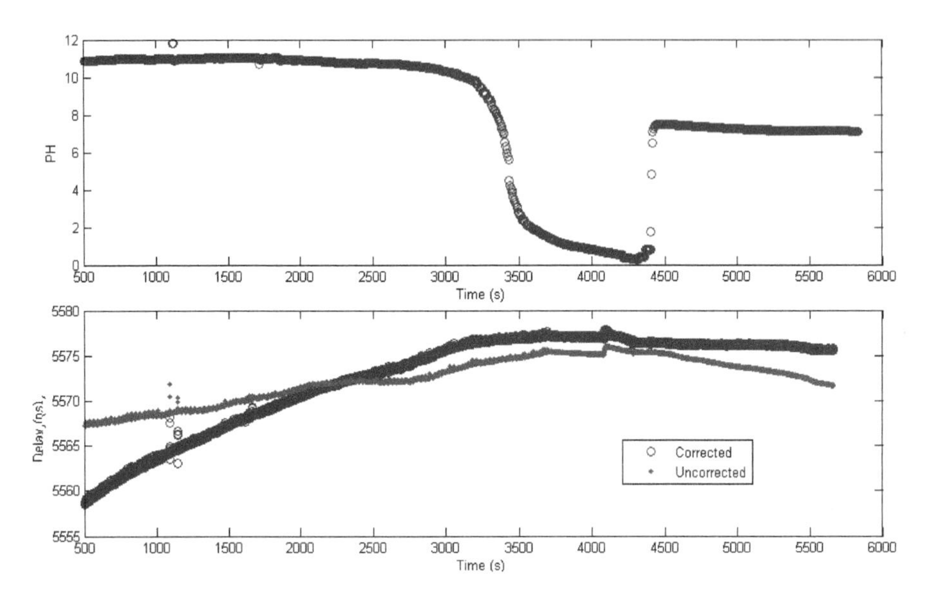

Fig. 4.11 Instrument readings for the second experiment.

influence. The overall conclusion from this dataset was that neither instrument exhibited sensitivity high enough to detect changes in chemical composition robustly at these concentrations.

In the second experiment, the concentration of NaOH in the analyte increased five-fold, to 0.1 M, as the standard solution was used undiluted; 1 mL of HCl at 0.1 M was added to the analyte every 30 seconds. Figure 4.11 shows that the endpoint (pH = 7) was reached after 110 additions of the acid. During this time, the measured ultrasound propagation delay notably increased and became nearly constant afterwards. Figure 4.11 displays the behavior of the ultrasound propagation delay during the entire experiment, using both raw (red) and corrected (blue) data.

In the third experiment 10 mL of HCl at 0.1 M were added to the solution (0.1 M NaOH) every 300 seconds. The recorded readings started to show steps after each addition (Figure 4.12). Nevertheless the behaviour of ultrasonic readings was found similar to the previous experiment for both raw and corrected propagation delays.

The comparison of the fast titrant release (adding a substantial amount of titrant at once, Figure 4.12) with its slow release (Figure 4.11) allowed establishing the ability of both the pH and ultrasound data for detecting small

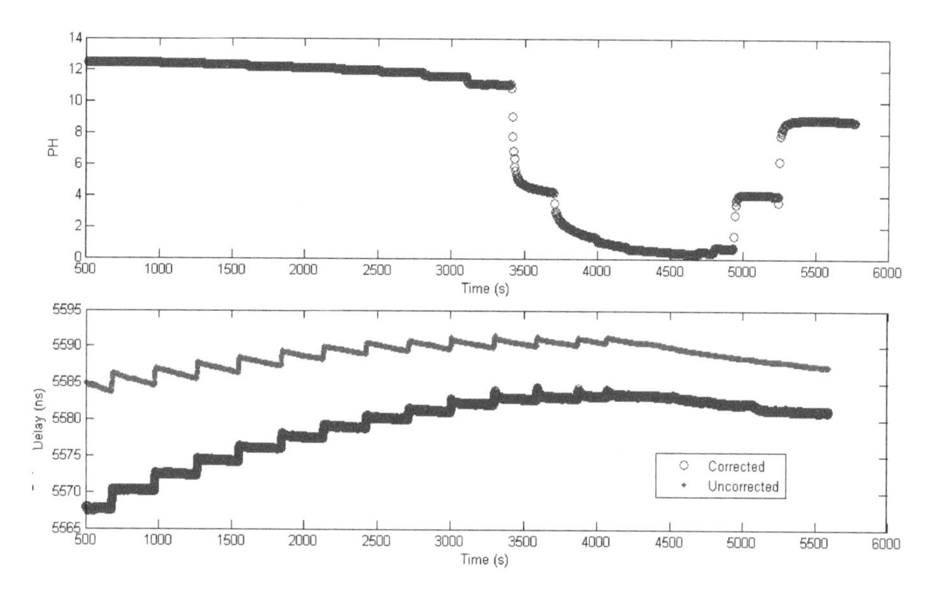

Fig. 4.12 Instrument readings for the third experiment.

continuous changes versus big sudden ones. Despite the similarity of both profiles, a closer look reveals the difference between both profiles. The ultrasonic and pH data collected during the second experiment around a single titrant drop are presented in Figure 4.13. It shows that whilst the pH readings did not notably change to indicate the addition of the titrant, the ultrasound propagation delay samples formed two distinct clusters before and after the addition. Therefore ultrasonic instrument was capable of detecting the change in the chemical composition at this time whilst the pH meter was not.

The achieved resolution for ultrasonic monitoring can be estimated by relating the weight of the chemical (HCl) released in one drop to the average total weight of the solution during the experiment:

$$\frac{0.1\,M \times 1\,ml \times 34.46\,g/mol}{\approx 180\,g} \approx 2 \times 10^{-5} = 20\,ppm \tag{4.10}$$

Therefore, the experimental sensitivity of the developed ultrasonic instrument to chemical changes improved ten-fold, compared to the sensitivity previously reported [21].

In summary, neither instrument was sensitive to step changes of the chemical composition of diluted (0.02 M) solutions at the titration rate of 0.5 mL/60 s. The developed instrument became sensitive to the changes at the titration

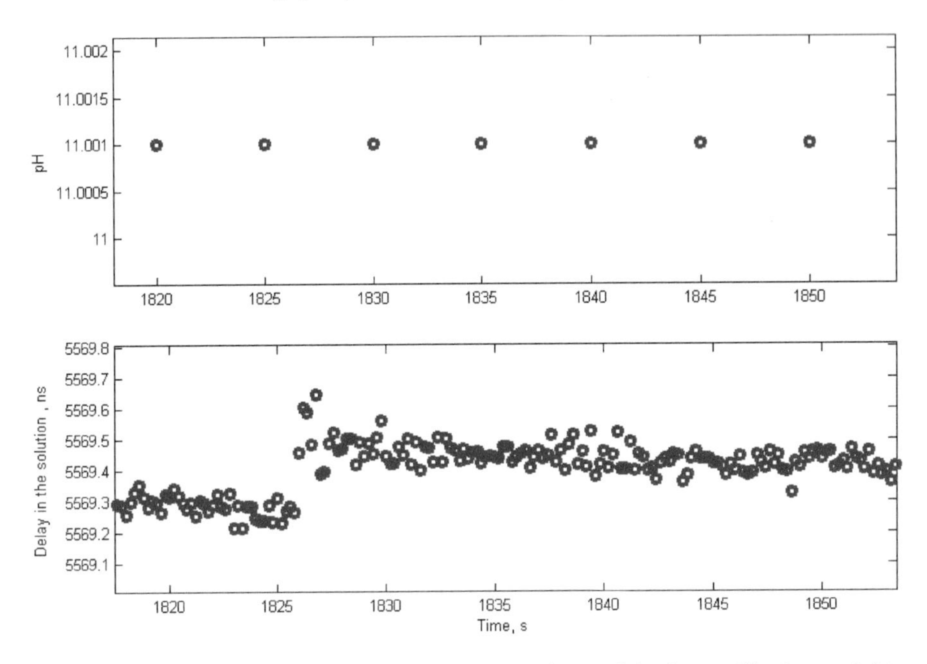

Fig. 4.13 pH data (top) were found not sensitive to slow releases of the titrant while ultrasound data (bottom) detected the change.

rate of 1 mL/30 s in standard (0.1 M) solution. Both instruments detected step changes at the titration rate of 10 mL/300 s in standard solutions.

4.7 Summary and Conclusions

The developed ultrasonic instrument integrated high resolution ultrasonic data acquisition instrumentation with fully automated data processing using a low cost FPGA and embedded soft core processor technology in order to achieve high measurement throughput for ultrasonic monitoring of chemical processes in real time. Data acquisition system featured on-the-fly averaging and also employed accurate interleaved sampling by dynamically applying phase shifts to the excitation pulse. The system utilized a 10 bit ADC clocked by a 50 MHz oscillator achieving data acquisition resolution of 16 bit $(10 + \log_2 64)$ at 1 GHz equivalent sampling frequency.

The operation of the instrument was verified by comparing the acquired waveforms with those obtained using a commercially available instrument.

The embedded processor estimated the ultrasound propagation delays using zero crossings. The processor removed the DC offset from the acquired waveforms and detected an ultrasonic pulse of interest using an adaptive threshold. It found two subsequent waveform samples with the opposite signs and estimated the zero crossing time instant by linear interpolation. Data acquisition, delay estimation, and communication took about 0.2 s overall, which was a tenfold improvement over the previous design. This increase was obtained by utilizing embedded processing of the acquired echo waveforms. Additionally the instrument no longer required a personal computer to calculate the propagation delays that were applied to the aqueous solution monitoring.

The developed instrument was used for monitoring chemical reactions, namely acid-base titrations, along with a commercially available pH meter. The reflector of the ultrasonic waves featured a water-filled cavity chemically isolated but thermally connected to the solution of interest. Measurements of the ultrasound propagation delays in this cavity enabled both temperature correction of the measured propagation delays in the solution and ultrasonic assessment of the solution's temperature. Analysis of experimental data led to the following observations:

- Ultrasonic temperature estimation could achieve much higher resolutions than the resolution of 0.1°C common to many commercially available instruments;
- The ultrasound propagation delays measured in the solution could show a substantial bias and misleading trends due to the temperature influence. These artifacts were largely removed by the applied temperature correction;
- Ultrasound monitoring allowed distinguishing among very close states of the solution that differ by around 20 ppm by weight, which was not possible with a conventional pH meter.

In summary, the combination of high accuracy waveform acquisition with embedded processing enabled the developed ultrasonic NDE instrument to achieve 5 measurements per second of the ultrasound propagation delay with the sensitivity of around 20 ppm per weight for chemical reaction monitoring.

References

[1] E. P. Papadakis, ed., *Ultrasonic Instruments and Devices II, Reference for Modem Techniques, and Instrumentation, Technology*, vol. XXIV of *Physical Acoustics*. Academic Press, 1999.

[2] F. Capote and M. Luque de Castro, *Analytical applications of ultrasound*, vol. 26 of *Techniques And Instrumentation In Analytical Chemistry*. Elsevier, 2006.

[3] "Thermal runaway, available online on http://tinyurl.com/y8u6uv3, accessed January 2011."

[4] A. Dukhin, P. Goetz, and B. Travers, "Use of ultrasound for characterizing dairy products," *Journal of dairy science*, vol. 88, pp. 1320–1334, April 2005.

[5] A. Kalashnikov and R. Challis, "Errors and uncertainties in the measurement of ultrasonic wave attenuation and phase velocity," *Ultrasonics, Ferroelectrics and Frequency Control, IEEE Transactions on*, vol. 52, pp. 1754–1768, oct. 2005.

[6] A. N. Kalashnikov, "Waveform measurement using synchronous digital averaging: Design principles of accurate instruments," *Measurement*, vol. 42, no. 1, pp. 18–27, 2009.

[7] V. Ivchenko, A. Kalashnikov, R. Challis, and B. Hayes-Gill, "High-speed digitizing of repetitive waveforms using accurate interleaved sampling," *Instrumentation and Measurement, IEEE Transactions on*, vol. 56, pp. 1322–1328, Aug. 2007.

[8] A. Kalashnikov, V. Ivchenko, R. Challis, and B. Haves-Gill, "High-accuracy data acquisition architectures for ultrasonic imaging," *Ultrasonics, Ferroelectrics and Frequency Control, IEEE Transactions on*, vol. 54, pp. 1596–1605, Aug. 2007.

[9] W. Chen, *Low cost instrumentation for high temporal resolution ultrasonic NDE with applications to solid objects*. PhD thesis, Electrical and Electronic Engineering-University of Nottingham, 2010.

[10] A. S. Dukhin and P. J. Goetz, "Bulk viscosity and compressibility measurement using acoustic spectroscopy," *The Journal of Chemical Physics*, vol. 130, no. 12, p. 124519, 2009.

[11] E. Zorebski, M. Zorebski, and M. Gepert, "Ultrasonic absorption measurements by means of a megahertz - range measuring set," *J. Phys. IV France*, vol. 137, pp. 231–235, 2006.

[12] S. Hickey, S. A. Hagan, E. Kudryashov, and V. Buckin, "Analysis of phase diagram and microstructural transitions in an ethyl oleate/water/tween 80/span 20 microemulsion system using high-resolution ultrasonic spectroscopy," *International Journal of Pharmaceutics*, vol. 388, nos. 1-2, pp. 213–222, 2010.

[13] S. Hickey, M. J. Lawrence, S. A. Hagan, and V. Buckin, "Analysis of the phase diagram and microstructural transitions in phospholipid microemulsion systems using high-resolution ultrasonic spectroscopy," *Langmuir*, vol. 22, no. 13, pp. 5575–5583, 2006.

[14] A. Kalashnikov, K. Shafran, V. Ivchenko, R. Challis, and C. Perry, "In situ ultrasonic monitoring of aluminum ion hydrolysis in aqueous solutions: Instrumentation, techniques, and comparisons to ph-metry," *Instrumentation and Measurement, IEEE Transactions on*, vol. 56, pp. 1329–1339, Aug 2007.

[15] M. Mather, J. Crowe, S. Morgan, L. White, A. Kalashnikov, V. Ivchenko, S. Howdle, and K. Shakesheff, "Ultrasonic monitoring of foamed polymeric tissue scaffold fabrication," *Journal of Materials Science: Materials in Medicine*, vol. 19, pp. 3071–3080, 2008. 10.1007/s10856-008-3445-y.

[16] E. B. M. Jager, U. Kaatze and V. Buckin, "New capabilities of high-resolution ultrasonic spectroscopy: titration analysis," *Spectroscopy*, 5, 2005.

[17] F. Priego-Capote and M. D. L. de Castro, "Analytical uses of ultrasound: Ii. detectors and detection techniques," *TrAC Trends in Analytical Chemistry*, vol. 23, no. 10-11, pp. 829–838, 2004.

[18] G. Steiner and C. Deinhammer, "Ultrasonic time-of-flight techniques for monitoring multi-component processes," *Elektrotechnik und Informationstechnik*, vol. 126, pp. 200–205, 2009. 10.1007/s00502-009-0640-6.

[19] K. L. U. Kaatze and F. Eggers, "TOPICAL REVIEW: Ultrasonic velocity measurements in liquids with high resolution techniques, selected applications and perspectives," *Measurement Science and Technology*, vol. 19, pp. 062001–+, June 2008.

[20] T. g. S. o. s. i. p. w. J. Ablitt, "http://resource.npl.co.uk/acoustics/techguides/soundpurewater/," 3, 2011.

[21] A. Kalashnikov, V. Ivchenko, R. Challis, and A. Holmes, "Compensation for temperature variation in ultrasonic chemical process monitoring," in *Ultrasonics Symposium, 2005 IEEE*, vol. 2, pp. 1151–1154, Sept 2005.

[22] "Spartan-3E Starter board, available online http://tinyurl.com/65o6c2p, accessed January 2011."

[23] A. Afaneh, *Processing of ultrasonic signals at the place of acquisition using soft and embedded processor cores*. PhD thesis, University of Nottingham, 2011.

[24] A. Afaneh, H. Yin, and A. Kalashnikov, "Implementation of Accurate Frame Interleaved Sampling in a Low Cost FPGA-based Data Acquisition System," in *The 6th IEEE International Conference on Intelligent Data Acquisition and Advanced Computing Systems: Technology and Applications*, pp. 20–25, 2011.

[25] PicoTech, "Picoscope 5024." http://www.picotech.com/picoscope5000.html, Last accessed June, 2011.

[26] B. Barshan, "Fast processing techniques for accurate ultrasonic range measurements," *Measurement Science and Technology*, vol. 11, no. 1, p. 45, 2000.

[27] A. Hammad, A. Hafez, and M. T. Elewa, "A labview based experimental platform for ultrasonic range measurements," *ICGST International Journal on Digital Signal Processing, DSP*, vol. 6, pp. 1–8, 2006.

[28] G. Cote and M. Fox, "Comparison of zero crossing counter to fft spectrum of ultrasound doppler," *Biomedical Engineering, IEEE Transactions on*, vol. 35, no. 6, pp. 498 –502, 1988.

[29] V. A. Afaneh, S. Alzebda, and A. N. Kalashnikov, "Ultrasonic measurements of temperature in aqueous solutions: why and how," *Physics Research International*, vol. 2011, p. 10, 2011.

5

Synchronization of Distributed Systems using GPS

Jan Breuer[1], Blanka Čemusová[2], Jan Fischer[1],
Jaroslav Roztočil[1] and Vojtěch Vigner[1]

[1] *Czech Technical University in Prague, Faculty of Electrical Engineering, Czech Republic*
[2] *The Institute of Photonics and Electronics, Time and Frequency Department, Prague, Czech Republic*

Abstract

The synchronization of large systems is important for recording of dynamic actions and data acquisition of physical quantities over a wide geographical area. The chapter deals with synchronization of distributed measurement systems, especially synchronization using PTP (Precise Time Protocol, IEEE 1588). A practical case study presents a Master Clock module synchronized by GPS receiver. This module works as a high quality time base for PTP based system.

Keywords: synchronization, IEEE1588, GPS, time scale, event timestamping

Advanced Distributed Measuring Systems — Exhibits of Application, 95–120.

5.1 Introduction

The data acquisition (DAQ) systems are used for acquiring information about various objects and processes. They perform multichannel monitoring and recording of signals representing electrical or non-electrical quantities. There are available compact centralized systems for fast multichannel synchronized data acquisition (VXI, PXI). The synchronization and control of data samples acquisition and generation of output signals are the main properties of these systems which can be done due to existence of specialized buses, as RTSI (Real Time System Integration bus), PXI trigger bus (used in systems of NI-National Instruments).

In case the large objects or areas should be observed, the signals from sensors is not possible to convey directly to centralized DAQ system and distribution of tasks to smaller DAQ units is necessary. The transfer of information from these smaller units is done in most cases in digital form. This is a typical case of distributed data acquisition system. One of the important problems arising in distributed DAQ system is a delay occurring between samples of data in different parts of distributed DAQ system. When signals representing behaviour of objects or processes are changing relatively slowly (e.g., signals corresponding to temperature, pressure, position) the delay between data samples taken from different places of an object is not critical. On the contrary, when rapidly changing quantities have to be observed and further data processed in order to determine their mutual relations and correlations, the deviation between time of samples of acquisition is of crucial importance.

In order to ensure the equality of sampling time instants in different parts of object, the problem of synchronization of units in distributed DAQ system has to be solved. The electrical power networks and analysis of it‘s dynamic behaviour (e.g., changes of load, failures) can be used as an example of large object [3]. Usually it is required to observe the behaviour of network in critical or ultimate states and to learn it‘s response to failures. Then electrical signals from various points of network should be acquired. At the same time it is necessary to match correctly time instants of sampling. In distributed dynamic system dedicated for multichannel data acquisition the synchronicity of sampling in individual channels plays very important role. Otherwise it is impossible to find mutual dependence and relation of phenomena in object under observation. Another typical example of large system occurs in seismology where large arrays of seismic sensors are employed and mutual relation between seismic waves in individual locations is searched.

5.2 Synchronization

Besides the measurement the distributed DAQ system can be also utilized for multichannel synchronous generation of control signals and driving signals for actuators. This is the case of large technological complexes where it is required to control simultaneous run of electrical machines. Here securing of synchronous multichannel signal sampling and generation of control signals by individual blocks of distributed DAQ system plays the key role.

On Figure 5.1 are depicted pulses of two channels (A, B) which can be used for sampling and for generation of control signals as well. These pulses are derived from internal time basis of control unit of individual channels. The periods of pulses are designated as T_A, T_B. The time shift is Δt_{1AB}. The equality $T_A = T_B$ is required, which means the frequencies f_A, f_B of control pulses are the same, i.e., $f_A = f_B$. The securing of identical frequency of control pulses is designated by word syntonization. In case of syntonization the sampling period will be also identical, nevertheless the instants of sampling might be shifted in time. Identical sampling periods are not satisfactory for purposes of data acquisition. In ideal case equality $\Delta t_{1AB} = 0$ should be valid representing the fulfillment of condition of individual channels synchronization.

In large distributed DAQ system the control of sampling by sending instantaneous commands from central control unit is not feasible due to differences and variation of command transfer time. The basic idea of this problem solution is to use unified time method for planning the time instants of sampling. The realization of this approach expects the existence of local clocks in each system unit. The local clocks should generate time scale coherent with reference time

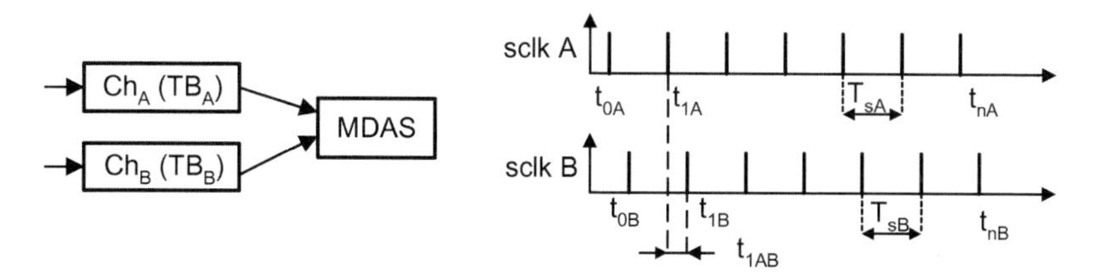

Fig. 5.1 Control pulses of two channels.

scale generated by system master clock. The main task is to secure coherence of time scales which can be, depending on distributed system dimensions and requirements on degree of synchronization, done in several different ways.

5.2.1 Synchronization of Time Basis

Independent identical generators of time basis in portable form having sufficient stability could be used for synchronization of individual points of distributed system. When their mutual synchronization is feasible, then the time information could be transferred to different places of system. Here they would serve as independent local time basis until the time of next synchronization.

Provided that planning of sampling instants in DAQ system would be done in exactly defined time points of unified time scale. This approach requires minimization of frequency deviations of individual time basis. The divergence of frequencies of individual basis should be so small that for duration of experiment and data acquisition the difference of time scales should fulfill inequality $\Delta t_{1AB} \ll T_A, T_B$. It means that deviation of the time of sampling in both channels should be substantially smaller than sampling period.

Two different generators have in long time interval always a different frequency of pulses. Due to changes of temperature and other factors, time fluctuations of clock frequency always occur. In further derivation the constant frequency f_A respectively f_B of clock generators for time between synchronization T_{sync} will be assumed for simplicity. On Figure 5.2 is depicted the case of signals from two generators *sclkA* and *sclkB* derived from corresponding time basis TB_A, TB_B, which were mutually synchronized at the beginning. Due to the difference of period T_A and T_B, pulses *sclkA* and *sclkB*, the time is shifted by deviation Δt_{AB} which with time increases T_{sync} elapsed from last synchronization. The value of Δt_{AB} could be calculated from relation (5.1),

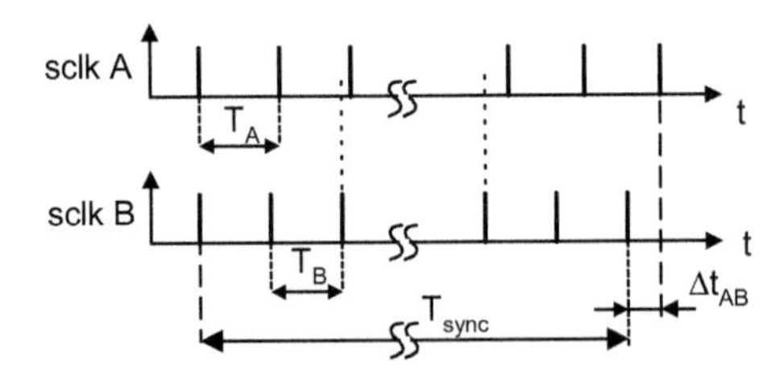

Fig. 5.2 Time basis with different periods.

where f_A respectively f_B are corresponding frequencies of pulses and n is number of pulses *sclkB* generated from last synchronization.

$$\Delta t_{AB} = n(T_A - T_B) = \frac{T_{\text{sync}}}{T_B}(T_A - T_B) = T_{\text{sync}} \cdot f_B\left(\frac{1}{f_A} - \frac{1}{f_B}\right) \quad (5.1)$$

$$\Delta t_{AB} = T_{\text{sync}}\frac{f_B - f_A}{f_A} = T_{\text{sync}} \cdot \delta_{fAB} \quad (5.2)$$

From relation (5.2) results that the value of time shift Δt_{AB} is directly proportional to time T_{sync} and relative deviation of frequencies — fractional offset δ_{fAB}. In case of two oscillators with frequencies $f_A = 9\,999\,999$ Hz and $f_B = 10\,000\,000$ Hz from which *sclkA* and *sclkB* will be derived, the value of $\Delta t_{AB} = 0.1$ ms for the time $T_{\text{sync}} = 1000$ s. It means that in case of DAQ systems utilizing these oscillators for generation of sampling pulses with $f_G = 1$ kHz and $T_G = 1$ ms will be for 1000 seconds deviation Δt_{AB} equal to 10% of period T_G of these pulses. This fact can be described by relation (5.3), expressing time shift related to the period of sampling pulses.

$$\frac{\Delta t_{AB}}{T_G} = T_{\text{sync}} \cdot f_G \cdot \frac{f_B - f_A}{f_A} = T_{\text{sync}} \cdot f_G \cdot \delta_{fAB} \quad (5.3)$$

The Equation (5.3) can be modified and used in form (5.4) for expressing time shift by number of periods of pulses *sclkA*. Lets suppose that nominal frequency of oscillator should be f_{AN}, but actual frequency is f_A. Then δ_{fA} is relative deviation of frequency f_A, from nominal value f_{AN}.

$$\delta_{fA} = \frac{f_A - f_{AN}}{f_{AN}} = \frac{\Delta f_A}{f_{AN}} \quad (5.4)$$

When clock signal *sclkA* with frequency f_A having deviation δ_{fA} is used for time base control, then the reading of this time base due to error of frequency will (after time T_{sync} when process of synchronization begins) differ according to (5.5) by the value m_A period *sclkA*.

$$m_A = \frac{\Delta t_A}{T_A} = T_{\text{sync}} \cdot f_A \cdot \delta_{fA} \quad (5.5)$$

For example in case of synchronization with interval $T_{\text{sync}} = 1$ s and $f_{AN} = 10\,000\,000$ Hz and $\delta_{fA} = 10^{-6}$, the shift will be equal $m_A = 10$. Thus between two subsequent synchronizations the content of counter will increase by 10 with respect to correct value (corresponds to time error of 1 μs).

So it is clear that for high requirements on synchronicity of distributed systems it is necessary to synchronize quartz controlled oscillators. The synchronization is performed by periodical introducing of local time base to state coherent with master clock of system. This process of introducing is done on the basis of synchronizing pulse or arrival time mark. The synchronization with exactly given period could be also used for syntonization. According to interval of synchronizing pulses or marks, the arrival numerically controlled generator of local time base is trimmed.

5.2.2 Synchronization by Means of Communication Channel

The communication channel which is used in distributed system for data transfer could be utilized also for synchronization of local time basis. As the channel is ment to be primarily for data transfer, it might not satisfy requirements of speed from the point of view of delay which occurs during information transfer. Nevertheless there is a chance to make use of it by determination of the channel delay using specialized autonomous form of communication. Knowing the delay, it is then easy to correct it. This represents basic idea of exploitation of interface Ethernet with protocol PTP IEEE 1588. With increasing dimensions of distributed system, where information passes across several communication blocks, the uncertainty of network parameters specification increases and as a consequence increases also uncertainty of delay correction and decreases the accuracy of synchronization.

5.2.3 Synchronization by Independent Radio Channel

In large distributed systems it is recommended to use communication channel only for allocation of requirements for data acquisition or actions and to perform synchronization by different means. Thus it is necessary to find alternative independent communication channel by means of which it will be feasible to compare local time basis of individual parts of distributed system. Here the utilization of radio channel seems to be a reasonable offer. The well known principle of clock synchronization exploits LW radio transmitter DCF77 which broadcasts time marks on frequency 77.5 kHz. Unfortunately due to undesired influences of ionosphere and other disturbing effects occurring during electromagnetic waves propagation, the uncertainties of time instant determination does not allow to use this approach for more demanding applications. The better properties can be reached by radio channels working in the range of UHF.

At the present time the exploitation of satellite navigational system, in particular GPS, represents substantially better variant of this approach. By means of GPS signals the local time basis of distributed system can be synchronized to absolute time UTC or to GPS time. In this case a communication channel of system is used only for control of data acquisition process, the transfer of results and overall activity control.

5.2.4 General Requirements for Local Oscillators from the Point of View of Synchronization

In individual blocks of DAQ system the quartz oscillators with variety of frequencies can be utilized. From the output signals of oscillators, the reference pulses of system clock by means of different circuits e.g., NCO can determine local time scale. By means of synchronization and using feedback concepts the NCO can be trimmed so that coherency of individual time scale is reached. By this approach the basic influence of basic deviation of quartz oscillator frequency from nominal value is eliminated. But the fluctuations of oscillator frequency remains. It can be stated generally that it is more advantageous to use high quality stable oscillator even if it has certain deviation from nominal frequency than oscillator having mean value of frequency close to nominal one but suffering from short time changes which must be corrected more frequent synchronization.

For reaching the best possible quality the inherent circuits of oscillator should be separated from microcontroller circuitry used in DAQ. The experiments show that autonomous oscillator, eventually with micro-thermostat, have better properties (in terms of higher stability and interval of synchronization T_{sync}) than standard quartz resonator linked to internal circuits of microcontroller.

5.3 Synchronization Over Ethernet

NTP For time synchronization over the Internet the Network Time Protocol (NTP) is widely used. The original protocol proposal was created by David L. Mills from University of Delaware. The present version of it, i.e., NTPv4 is described in document RFC 5905 [12]. The protocol employs an algorithm for determination of estimation of exact time scale from several sources and is capable to eliminate the erroneous sources of time. It contains also the filtration algorithm for improving estimation from several synchronization messages.

The protocol uses UTC time scale. Information about time zones and switching between summer and winter time is out of NTP scope and is not available. In order to simplify implementation, the protocol SNTP was created which does not require preservation of system status for longer time. This protocol can be found e.g., in embedded systems.

The standard implementation of protocol has software character and allows to reach accuracy of 100 μs in local networks. The accuracy in Internet is worse and corresponds to several tens of ms.

PTP Precision Time Protocol (PTP) is defined by standard IEEE1588 [7] and is designed for precise time synchronization. It uses synchronization based on master-slave communication. The Master sends periodically messages containing time mark and by this marks slave devices are synchronized. In order to increase accuracy, the time mark is not sent in synchronization message but in a message, which immediately follows. In case when HW support exists, the time mark can be inserted in currently transmitted packet. For the purpose of the delay in transmission path correction, the slave device asks for the length of transmission path.

The PTP protocol is designated mostly for local networks and is most often used in Ethernet type networks. The multicast is used preferably for communication, but application of unicast is also defined. The protocol is under permanent development and it's actualization containing new developments was issued in 2008. The protocol describes auto-configuration properties, known as Best Master Algorithm (BMC), which enables the selection of best master unit, to which are then all remaining units synchronized.

Using purely software implementation only moderate improvement of time synchronization can be reached, the synchronization error might have the value in the range from 10 to 100 μs.

Precise Time Protocol is meant to be for cases when accuracy of NTP is not sufficient and implementation of e.g., GPS is not suitable or possible.

SyncE It is expected that TDM network technology will be replaced by synchronized packet networks. Due to this assumption the ITU-T created specification for synchronous Ethernet, which enables the distribution of frequency across Ethernet physical layer. The standard is described in specification ITU-T G8261 [9]. Respecting the original role of synchronous Ethernet in telecommunications, the standard is designed to be compatible with SHD-based networks. The devices designated for application on synchronous Ethernet are defined in ITU-T G8262 [10]. Unlike the previous possibilities based on packet exchange, serves only for syntonization.

As it can be deduced from previous facts, it is clear that combination of SyncE and IEEE 1588 is possible. SyncE is used for syntonization and IEEE 1588 for distribution Time of Day (ToD).

5.4 Precise Time Protocol (IEEE 1588)

5.4.1 The Description of Protocol Properties

Precise Time Protocol (PTP) serves for synchronization of units on packet network. The protocol is based on hierarchic architecture Master-Slave. The hierarchy is composed automatically by means of Best Master Clock algorithm, which uniquely determines the best clock in the system to which all other units are synchronized. In original version of protocol the master periodically emitted synchronizing messages by means of multicast UDP packets. Respecting the fact that the multicast limits the whole protocol only to local network, the following version was complemented by the possibility of synchronization by means of unicast or other layers. The possibility to exchange packets directly on second layer was also completed. The slave units ask for information about the delay on transmission path and on the base of this knowledge they correct the delay.

Originally the protocol was designed for time synchronization of measuring systems, but later on it was accepted also in other branches, i.e., for power distribution, telecommunication and automotion. The protocol can be used not only for time synchronization but also for syntonization, i.e., for linking to frequency of master devices.

5.4.2 Time Properties

The crucial problem of precise synchronization is acquisition of timestamps. The most accurate way of timestamping is direct application of hardware, i.e.,

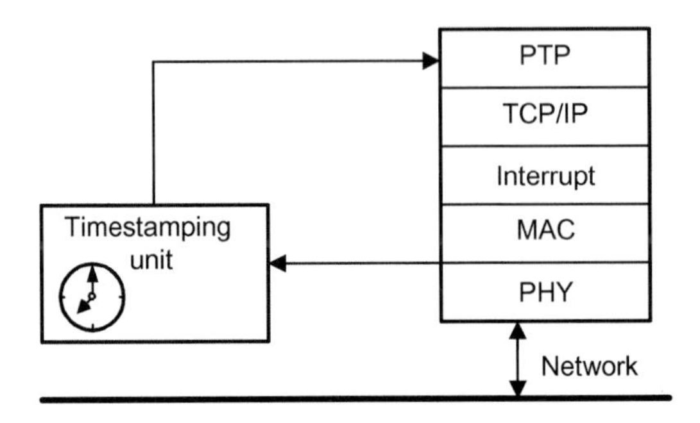

Fig. 5.3 Timestamp acquisition.

using physical layer. The closer is process of timestamping to physical layer, the less uncertainties threaten the result accuracy (see Figure 5.3).

The acquisition of timestamps in application layer is also possible at the price of uncertainty increase. The packet has to pass across hardware interface to protocol stacks up to application and anywhere on a way unexpected delay might occur.

It is important to save timestamps in event packets. The Sync packets followed by Follow Up packets serve for basic syntonization (see Figure 5.4). After arrival of at least two Sync packets, the derivation of frequency and linking to it is possible. The arrival of larger number of Sync packets increases the accuracy of master clock frequency estimation. Naturally together with Sync packets arrives also time information thus the slave node is even capable of time synchronization. But this information is loaded by error of time caused by Sync message transfer delay. In order to enable calculation of path delay slave sends query to the master asking for the length of transfer path. Master node responds by timestamp designating time of this query arrived.

The protocol contains an important mechanism serving for determination of delay on transfer path. The typical value of an ordinary switch delay is about 10 μs. In case when error of delay of transfer across the switch is not compensated, for the phase divergence of master and slave node would reach at least the mentioned value of 10 μs. Naturally, in more complicated topology the total delay of message further increases.

The measurement of the length of transfer path is performed by using couple of messages. In case when node is syntonized, one arriving Sync message is exploited. The time of departure of this message is t_1 and time of message

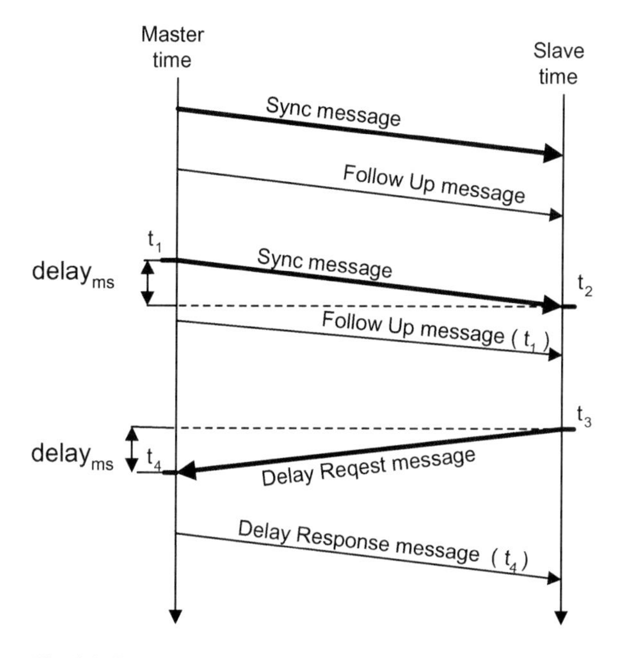

Fig. 5.4 The principle of synchronization — exchange of packets.

arrival is t_2. The delay of incoming packet is calculated as

$$delay_{ms} = t_2 - t_1 \tag{5.6}$$

The slave node sends Delay Request message in time t_3 and master sends back timestamp t_4 of this message acceptance. The delay of outgoing packet is then calculated as

$$delay_{sm} = t_4 - t_3 \tag{5.7}$$

Finally from these two values the mean time of delay on transfer path is calculated

$$meanPathDelay = \frac{delay_{sm} + delay_{ms}}{2} \tag{5.8}$$

The condition for correct calculation is the equality of time of message transfer in both directions. It is not possible to find asymmetry of transfer path by means of this algorithm. If asymmetry of transfer path exists, there are no means how to measure it and thus it manifests as the error of synchronization. In case when asymmetry is known, it is possible to compensate it manually.

There are some special extensions of protocol allowing to measure assymetry by sending special message and by application of special physical layers.

In case of direct master — slave connection this concept is satisfactory. When more complicated network composed from several network units not supporting PTP is used, there is no other way of calculation of time delay on transfer path. The method of delay time discovery described above is designated as End to End (E2E). But this approach is not suitable in situation when reconfiguration of network occurs. The length of total transfer path from master to slave is found by default (see Figure 5.5a). In case of master failure the BMC algorithm must be used for reconfiguration of whole network to another master and moreover each slave must again measure the total length of transfer up to the new master (see Figure 5.5b).

The increase of reconfiguration speed can be achieved by measurement of delay using the Peer to Peer (P2P) algorithm. In this configuration is necessary that network will support PTP. Each device measures delay on all its ports to closest device. This delay is not measured across whole trace master-slave, but for example only across trace from slave to transparent clock (see Fig. 5.6a).

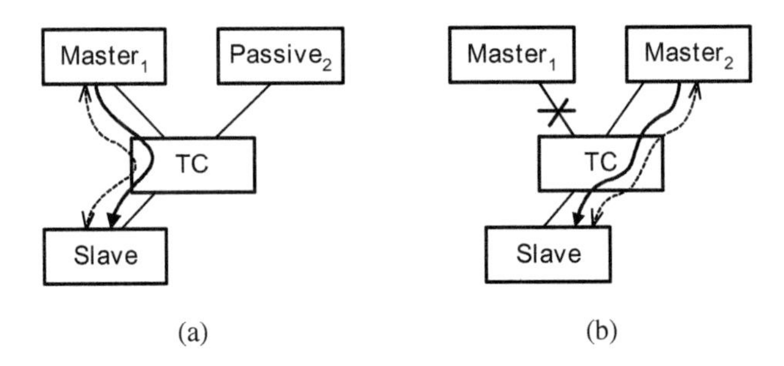

Fig. 5.5 E2E and reconfiguration.

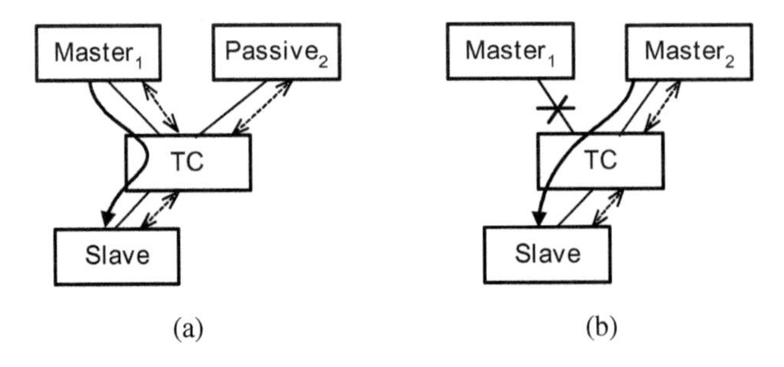

Fig. 5.6 P2P and reconfiguration.

More over each unit of network is able to measure time interval for which packet stay in it. The total time spent on transmission path is calculated from partial time delays. In case of malfunction of master, the new one is chosen using BMC algorithm. However the difference between P2P and E2E lies in fact, that it is not necessary to measure again the delay on transmission path, because it has been already measured (see Fig. 5.6b). Thus it is necessary to add partial contributions to a new master device. Using this approach leads to the decrease of the time interval necessary for reconfiguration in case when malfunction occurs.

5.4.3 Topological Properties

The basic components of device topology are Ordinary Clock (OC), Grand Master Clock (GMC), Boundary Clock (BC) and Transparent Clock (TC) (see Figure 5.7).

Ordinary Clock can act as Master or Slave. In case when network is composed from OC only, the best clock would be chosen as Master and other devices would be synchronized to it.

Grand Master Clock is a device which is either sufficiently stable for a given purpose or is synchronized from other reference which does not support PTP. As an example could serve GPS receiver. GMC works mostly in two regimes, master and passive one. In master regime it works during the time interval when there is no better GMC in the network. In passive regime it only watches whether in network exists better master. If this is not a case, it takes the role of reserve clock. Provided the reserve clock is out of order, the role of master would be taken by one of connected OC.

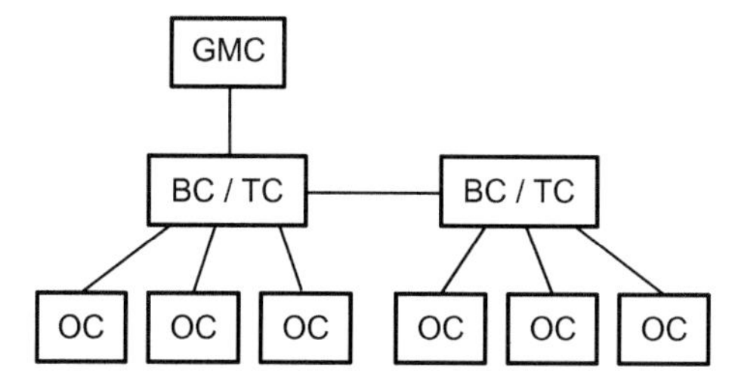

Fig. 5.7 Typical hierarchy of units at synchronization by means of IEEE 1588.

Boundary Clock is a multiport device and on all its ports behaves as an ordinary switch. But in case PTP packet appears, the relevant port catches it. The behaviour of the individual ports of device depends on regime in which they occur. In passive regime the port only follows the incoming packets. In slave regime the port is synchronized with other units of network. BC can have only one port in slave status. The rest of ports are in status of master. The units connected to these ports are synchronized by them.

Transparent Clock is a multiport device and similarly as BC behaves as an ordinary switch. The only difference is it‘s behaviour with respect to PTP packets. The BC must be synchronized which is it‘s main disadvantage. The synchronization might not be sufficiently accurate and thus device introduces some error to all subsidiaries. In case there are more BC connected in series, the error accumulates. TC works on other principle. For proper function only syntonization of TC is required. Event packet which passes across unit is marked at the input and output of device. Due to syntonization the time of transfer of packet across the device can be relatively exactly calculated. This time is sent out to the next packet. When more TC are connected in series, the accumulation of error is not so severe. BC accumulates errors of syntonization and synchronization whereas TC accumulates only error of syntonization, which is smaller.

Best Master Clock algorithm selects unique clock in system to which all other are synchronized. BMC algorithm is distributed in whole system and ensures the robustness of entire synchronization. The goal of algorithm is to find the best clock in entire system. If all qualitative comparisons fail, one device is chosen as master according to its identification.

The algorithm is based on expiration of timeout from last Announce Message carrying information from actual master device. Expiring of timeout means absence of master in network leading consequently to switching of some unit to regime master and sending it‘s Announce Message. Since message is sent by multicast it reaches all nodes. In time instant of receiving foreign message, the node compares it with it‘s abilities. If it is evaluated as better master it sends it‘s own Announce Message.

Provided that present master accepts Announce Message from better master node, it switches to regime slave or passive depending on node setting.

The main parameters of comparison are following:

- *Priority* — manually set value of clock for highlighting the particular clock system
- *Quality* — this parameter includes information about the type of clock and source from which information about time is obtained. Synchronization by GPS has always better quality then free running quartz oscillator.
- *Variance* — is a value expressing the measured deviation of parameters with respect to reference clock
- *Identifier* — is an unique and universal identifier of a clock. In Ethernet network identifier is derived from MAC address of the network port

BMC algorithm continues to act until stable state is not reached. Then at this state one best master is chosen.

5.4.4 Implementation of Protocol in Ordinary Clock

The Protocol IEEE 1588 is defined as the state automaton. The most important states are Initializing, Master, Slave and Passive. The whole protocol accumulator might be implemented only by software means. Nevertheless for better precision the hardware support is more suitable. The example of software implementation is project PTPd [2].

Implementation by hardware support is possible thanks to specialized integrated circuits supporting IEEE 1588 protocol. By hardware support is understood mainly a possibility of saving timestamps of incoming and outgoing packets and also possibility to slightly modify internal clock. Moreover some components allow recording of timestamps accompanying external events, generation of pulses at precise time instants and generation of signal with frequency derived from synchronized clock.

Microcontroller STM32F107 from ST Microelectronics contains support for IEEE 1588 in MAC interface. Thanks to this fact the software PTP stack with hardware assistance can be utilized. By hardware assistance is understood a possibility to acquire an exact timestamp to each incoming and outgoing packet. Moreover Numerically Controlled Oscillator (NCO) is implemented in MAC interface (see Figure 5.8).

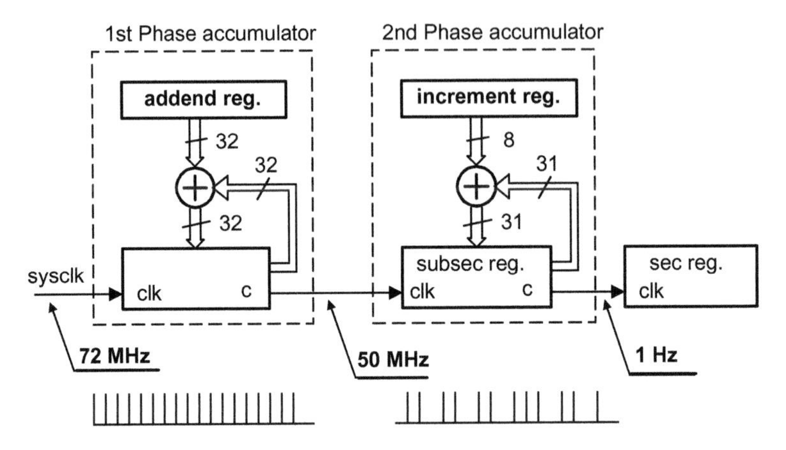

Fig. 5.8 Principle of digital clock adjustment in processor STM32F107.

Due to existence of NCO the implementation of system clock and its fine adjustment for exact timestamps generation is possible. Two phase accumulators are included in microcontroller STM32F107. The width of first phase accumulator is 32 bits and the value of Addend register is added to its content at each tick of system clock of processor. After reading of accumulator the pulse to further phase accumulator is generated. In this way the first accumulator enables to divide frequency of system clock by arbitrary value in the range from $2^{32}/(2^{32} - 1)$ up to $2^{32}/1$. The further phase accumulator adds at each input pulse a value from 8-bits Increment register to 31 bits‘ register. The combination of values of Addend and Increment registers is chosen in a way that sub-second part of full timestamp would appear directly in final register. After each overflow of this register the second part of timestamp register increments [11].

5.4.5 Realization on Platform with STM32F Connectivity Line

The implementation of protocol IEEE 1588 on platform with microcontroller STM32F Connectivity Line consists in utilization of freely acceptable protocol accumulator PTPd and supplementation of the function of this accumulator by the support of hardware timestamps. The final implementation is described in Application Note [5].

The measurement performed at Department of Measurement of Faculty of Electrical Engineering, CTU in Prague consisted on direct connection of

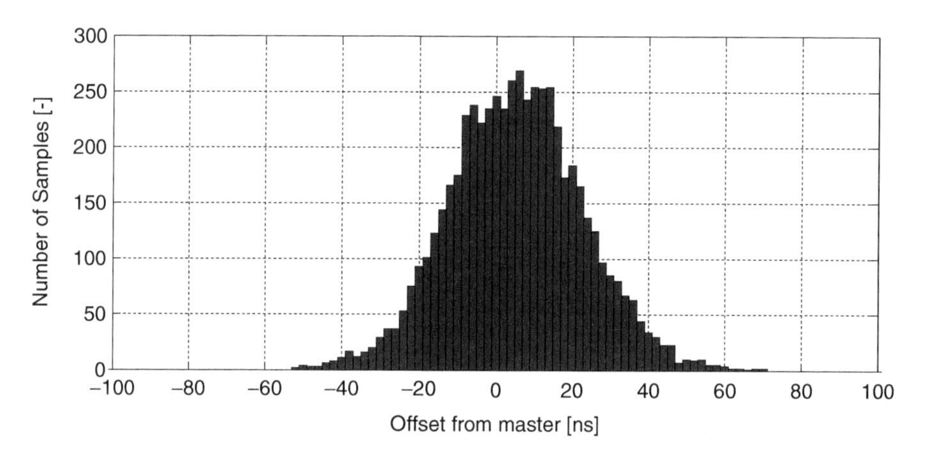

Fig. 5.9 Histogram of measured offset PPS — synchronization interval 1 s.

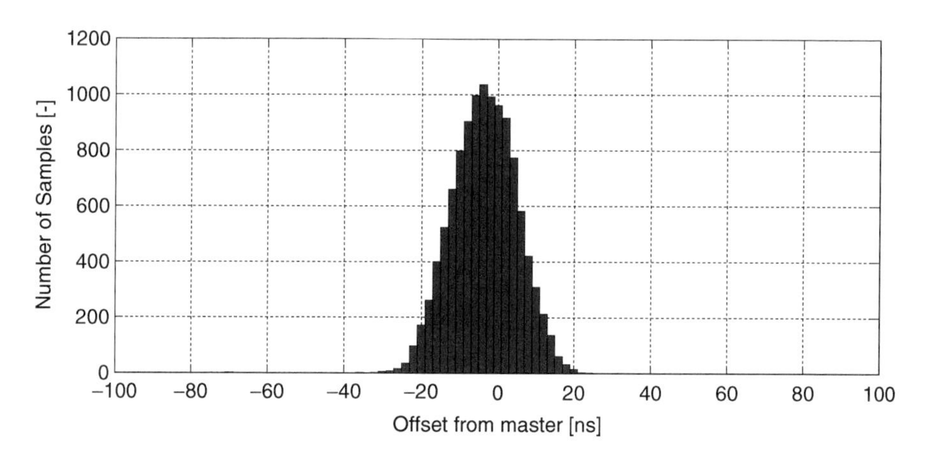

Fig. 5.10 Histogram of measured offset PPS — synchronization interval 0.125 s.

processor module and Meinberg Lantime M600 in function of master clock. The PPS signal was taken out from both units. The time delay of these signals was measured by universal time interval counter (SRS model SR620). The measurement was performed for synchronization interval 1 s and 0.125 s. The resultant histograms of measured offset are on Figure 5.9 resp. Figure 5.10.

5.5 GPS

Global Positioning System (GPS) is a navigation system available worldwide. It consists of 24 satellites in orbit around the Earth. The GPS satellites

Table 5.1. Typical uncenrtainties of GPS measurement techniques.

Technique	Timing Uncertainty 24 h, 2σ	Frequency Uncertainty 24 h, 2σ
One-Way	<20 ns	$< 2 \times 10^{-13}$
Single-Channel Common-View	$\cong$ 10 ns	$\cong 1 \times 10^{-13}$
Multi-Channel Common-View	<5 ns	$< 5 \times 10^{-14}$
Carrier-Phase Common-View	<500 ps	$< 5 \times 10^{-15}$

are controlled and operated by the United States Department of Defense (USDOD). Each satellite is equipped with atomic frequency standards referenced to Coordinated Universal Time (UTC) maintained by the United States Naval Observatory (USNO).

GPS enables worldwide continuous low-cost precise time transfer (see [14]). There are many types of GPS receivers differing in cost, design and accuracy, but most of them share several common features. Timing GPS Receivers are widely used for time synchronization and frequency calibration. These receivers provide a 1 pulse per second (PPS) timing output, and, in some cases, standard frequencies such as 1, 5, and 10 MHz. The 1 PPS output of low-cost Timing GPS Receivers can easily be synchronized to within 100 ns of UTC. This accuracy is usually sufficient for industrial applications.

Properly designed Precision Timing GPS Receivers can provide traceability to the national frequency standards (e.g., UTC (NIST), UTC (PTB), UTC (NPL), and so on).

There are three different types of GPS measurements used in time and frequency metrology: one-way, common-view, and carrier-phase. Typical uncertainties of GPS Measurement Techniques are presented in Table 5.1 (see [14]).

5.6 Device for Time Scale Generation and Event Timestamping

5.6.1 Objectives

Accurate time and frequency measurements, event timestamping and device time synchronization in distributed systems require a stable time scale. The use of precision oscillators is usually sufficient for short-term measurement. But the precision of time scale in case of long-term measurement decreases due to oscillator frequency shift and instabilities. This paragraph deals with combining the benefits of high-quality oscillator with good short-term stability and GPS receiver phase-locked to UTC scale. The goal is to create device which could be considered as a reference source of UTC. This combination enables to correct oscillator's error using the GPS PPS (Pulse Per Second) signal. Using timing GPS receiver generates PPS pulses rising edges of which corresponds to the UTC seconds marks. In case of the used ORCAM 30F GPS and uBlox LEA-6T receiver the accuracy of PPS pulses guaranteed by device manufacturer equals to ± 1 s resp. ± 0.1 s.

Many time standards based on crystal or rubidium oscillators can be found on market in various precisions classes and corresponding prices, see [1]. A device constructed at Department of Measurement FEE CTU in Prague (see [18]) is specifically designed for use in distributed systems. This device serves not only for time scale generation but also for event timestamping e.g., time identification of captured events generated by external sources. Thanks to this ability the device is suitable for time critical measurements.

5.6.2 Device Construction and Software

The device for time scale generation and event timestamping uses a FPGA Spartan3 from Xilinx and ATMEL's ATXmega microprocessor. These components are connected together using parallel interface (FPGA acts like a SRAM). This conception provides ability to separate design into two different parts. The subsystem (with 250 MHz inner clock) based on FPGA circuit and the control part containing microcontroller which controls the system programmed inside FPGA. A Matlab model of device was created in order to design the best deviation correction algorithm. Several types of regulators have been implemented for time deviation correction. The main problem is to suppress jitter of PPS signal from GPS receiver. A block diagram of prototype device (named Time-Base 1.0) used for current measurement is shown in Figure 5.11.

Reading and changing device parameters (such as deviation from synchronization PPS and correction value) is enabled by standard SRAM interface. This enables MCU to extend his memory space for e.g., program variables. Measured values are read-only and they are refreshed automatically. New measurement is reported by his own interrupted signal.

For the Time-Base 1.0 construction the MTI-Milliren Technologies-210 Series oven controlled quartz crystal oscillator (OCXO, see [15]) has been

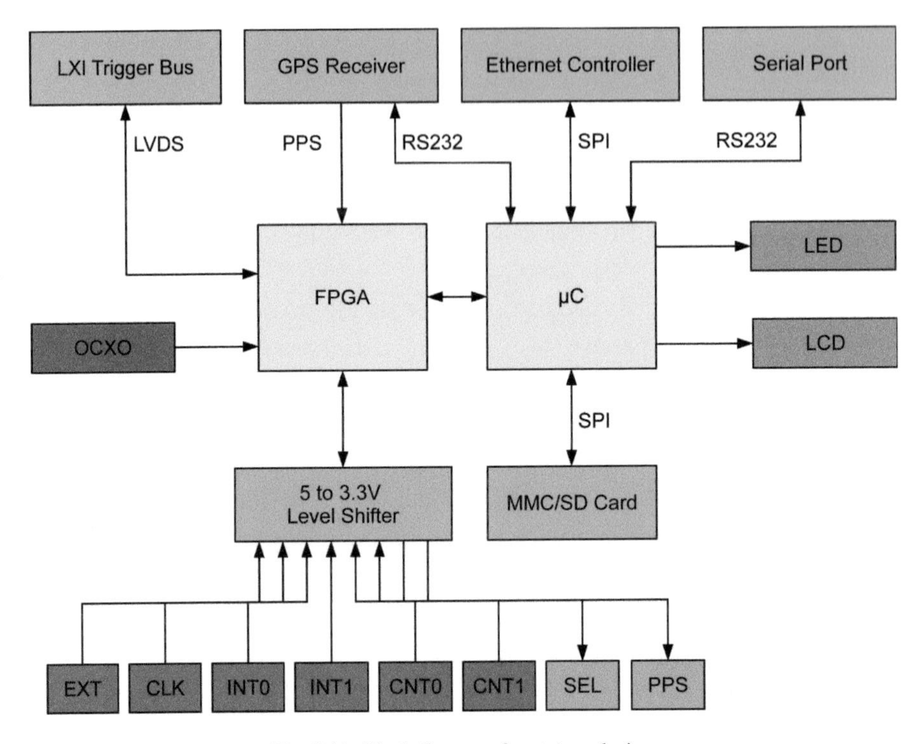

Fig. 5.11 Block diagram of prototype device.

selected. This type (MTI210) enables tuning of oscillator output frequency using analog voltage input. As it is described further this feature is not implemented in current solution.

Thanks to 250 MHz inner clock generated from external 10 MHz oscillator using PLL realized by FPGA, the resolution of time scale is 4 ns. So far the correction of time scale is based on simple digital frequency control. That means that the resolution of digital control equals to 1 Hz. The GPS receiver (integrated on the same board as the rest of the device) generates PPS signal with accuracy of 1 *s* with respect to UTC. In order to achieve a higher accuracy of time scale the time deviation between GPS PPS and device PPS signal is measured in longer time intervals (from 10 to 1000 seconds). This solution improves the accuracy without any side effect on regulation itself. Due to periodical changes of pre-scaler value used to generate PPS signal, all of the regulator variables (e.g., in PID regulator) have to be compensated for every control action (corresponding to measurement interval).

5.6.3 Measurement

This section contains description of applied measurement methods to confirm suitability and improvement of device construction. PPS signal generated by the Time-Base 1.0 was compared with PPS signal from rubidium frequency standard (SRS model FS725) synchronized by means of external GPS receiver ORCAM 20. Time delay between these two pulses was measured using universal time interval counter (SRS model SR620). This section contains description of applied measurement methods to confirm suitability and improvement of device construction.

5.6.4 Results

Figure 5.12 contains Allan deviation of synchronized and free-running time scale. The methods of measurement are described in [17] and [4]. Data were processed using Stable32 [16] and Matlab. Measured data show that the device is able to improve parameters of OCXO oscillator in short-term and long-term stability.

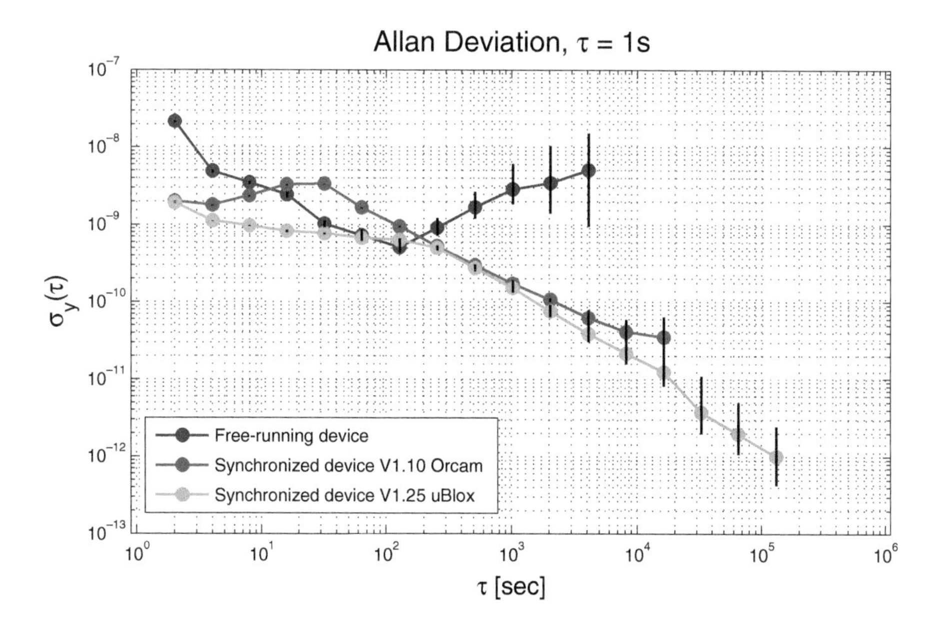

Fig. 5.12 PPS pulse stability of different settings and vesions of TimeBase.

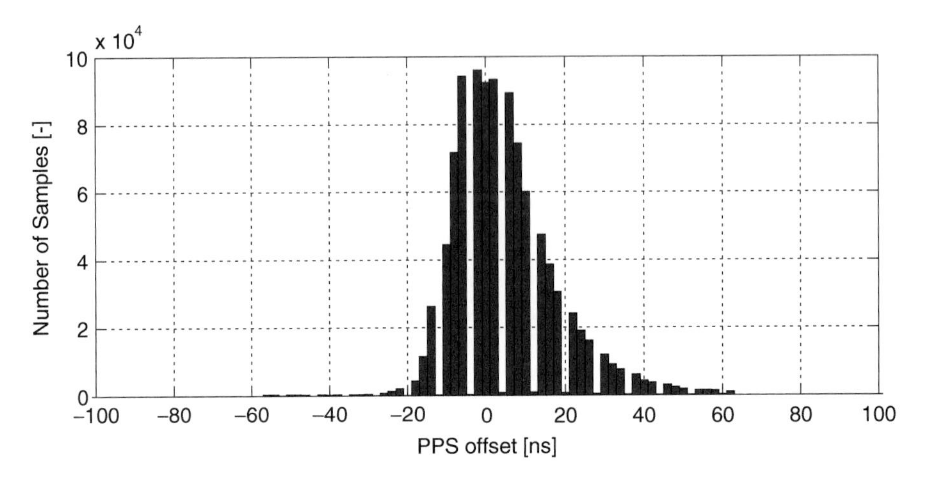

Fig. 5.13 Histogram of measured PPS offset.

Figure 5.12 also shows comparison of two different versions of synchronization algorithm. The version V1.10 Orcam (red) is the first version designed for device testing and the V1.25 uBlox (green) is final version which uses uBlox LEA-6T timing GPS receiver instead of ORCAM 30F GPS receiver.

Histogram of measured offset is in Figure 5.13 and represents normalized offset measured against reference PPS pulse.

5.7 Conclusion

This document presents basics methods to enable synchronization in large distributed systems. First is the implementation of IEEE 1588 protocol into the system, that presumes a bus connecting individual devices is capable of this implementation. So this method is suitable for local distributed systems. If distance between individual nodes is large (worldwide, statewide) then synchronizing individual nodes are more expensive or even impossible. This is where the other approach comes handy. That is the synchronization by electromagnetic waves (radio signal) from defined source. In this case using GPS signal.

Combination of these two methods allows us to create stable and synchronous large distributed system. Correct functionality of these constructed systems must be determined by long time measurements. Next major research is the influence of individual parts to stability and precision of whole system.

Acknowledgment

The authors would like to thank prof. Stanislav Ďaďo for his assistance, helpful comments and suggestions.

Terminology; Characterization of Oscillators and Clocks

In the further text the basic terms and definitions used for characterisation and measurement of frequency and time devices will be reviewed. The detail information can be found in IEEE 1139 (IEEE Standard Definitions of Physical Quantities for Fundamental Frequency and Time Metrology-Random instabilities [6]). The problems of synchronization of transmission systems, digital systems and networks is covered in standard ITU-T G.180 [8]

Definitions Related to Clock Characterization

Time scale: A system of unambiguous ordering of events. A time scale allows to date any event. Time scales use oscillators to define the length of the second, which is the standard unit of time interval. Modern standardized time scales TAI and UTC (see [13]) define the second based on a property of the cesium atom, earlier time scales were based on astronomical observations (measurement of the Earth's rotation).

Time Interval: The elapsed time between two events.

Clock: A device that generates periodic, accurately spaced signals used for timing applications. A clock consists of three basic parts: an oscillator, a counter and a means of displaying or recording the results. In some cases the clocks are defined as (see [8]): an equipment that provides a timing signal.

Timing signal: Timing signal s(t) is represented by:

$$s(t) = A \cdot \sin\Phi(t) \tag{5.9}$$

where A is a constant amplitude coefficient (assuming that the amplitude fluctuations are negligible) and $\Phi(t)$ is the total instantaneous phase of an actual timing signal.

$\Phi(t)$ is modelled as:

$$\Phi(t) = \Phi_0 + 2\pi\nu_{\text{nom}}(1 + y_0)t + \pi D\nu_{\text{nom}}t^2 + \varphi(t) \tag{5.10}$$

where Φ_0 is the initial phase offset, ν_0 is the nominal frequency, y_0 is the fractional frequency deviation from the nominal value ν_{nom}, D is the linear fractional frequency drift rate (aging effects) and $\varphi(t)$ is the random phase deviation component.

The total phase $\Phi_{id}(t)$ of an ideal timing signal is modelled as:

$$\Phi_{id}(t) = 2\pi \nu_{\text{nom}} t \tag{5.11}$$

where ν_{nom} is the nominal frequency of the reference oscillator.

Fractional frequency deviation (offset, departure) The difference between the actual frequency of a signal and a specified nominal frequency, divided by the nominal frequency. The fractional frequency deviation $y(t)$ can be expressed as:

$$y(t) = \frac{\nu(t) - \nu_{\text{nom}}}{\nu_{\text{nom}}} \tag{5.12}$$

where $\nu(t)$ is the instantaneous frequency (time derivative of the phase divided by 2π).

Time function: The time of a clock is the measure of ideal time t as provided by that clock. Time function $T(t)$ is defined as:

$$T(t) = \frac{\Phi(t)}{2\pi \nu_{\text{nom}}} \tag{5.13}$$

Time error function: Time error is the difference between the time of a clock generating time $T(t)$ and a reference clock generating time $T_{\text{ref}}(t)$. Time error function is defined as:

$$x(t) = T(t) - T_{\text{ref}}(t) \tag{5.14}$$

Based on the definition of time error and the above model of $\Phi(t)$, the the time error may be represented as

$$x(t) = x_0 + y_0 t + \frac{D}{2} t^2 + \frac{\varphi(t)}{2\pi \nu_{\text{nom}}} \tag{5.15}$$

where x_0 is the time offset (the origin of the time scale relative to the origin of the reference time scale, y_0 is the fractional frequency deviation from the nominal value ν_{nom}, $\varphi(t)$ is the random phase deviations of the measured oscillator.

References

[1] David W. Allan. Statistics of atomic frequency standards. *Proceedings of the IEEE*, vol. 54, no. 2, pp. 221–230, 1966.

[2] Kendall Correll and Nick Barendt. Design considerations for software only implementations of the ieee 1588 precision time protocol. In *Conference on IEEE 1588 Standard for a Precision Clock Synchronization Protocol for Networked Measurement and Control Systems*, 2006.

[3] Loredana Cristaldi, Alessandro Ferrero, Carlo Muscas, Simona Salicone, and Roberto Tinarelli. The impact of internet transmission on the uncertainty in the electric power quality estimation by means of a distributed measurement system. *Instrumentation and Measurement, IEEE Transactions on*, vol. 52, no. 4, pp. 1073–1078, Aug. 2003.

[4] Leonard S. Cutler and Charles L. Searle. Some aspects of the theory and measurement of frequency fluctuations in frequency standards. *Proceedings of the IEEE*, vol. 54, no. 2, pp. 136–154, 1966.

[5] *AN3411 Application Note. IEEE 1588 precision time protocol demonstration for STM32F107 connectivity line microcontroller.* ST Microelectronics, Doc ID 018905 Rev 1, Jul 2011.

[6] *IEEE Std 1139-1999, IEEE Standard Definitions of Physical Quantities for Fundamental Frequency and Time Metrology — Random Instabilities.* The Institute of Electrical and Electronics Engineers, Inc., New York, 1999.

[7] *IEEE Std 1588-2008, IEEE Standard for a Precision Clock Synchronization Protocol for Networked Measurement and Control Systems.* The Institute of Electrical and Electronics Engineers, Inc., New York, 2008.

[8] *ITU-T G.810. Definitions and Terminology for Synchronization Networks.* International Telecommunication Union, Geneva, 1996.

[9] *ITU-T G.8261: Timing and synchronization aspects in packet networks.* International Telecommunication Union, Geneva, 2008.

[10] *ITU-T G.8262: Timing characteristics of a synchronous Ethernet equipment slave clock.* International Telecommunication Union, Geneva, 2010.

[11] *RM0008 Reference manual: STM32F101xx, STM32F102xx, STM32F103xx, STM32F105xx and STM32F107xx advanced ARM-based 32-bit MCUs.* S T Microelectronics, Doc ID 13902 Rev 14, Oct 2011.

[12] David Mills *et al.* Network Time Protocol Version 4: Protocol and Algorithms Specification. RFC 5905 (Proposed Standard), June 2010.

[13] Judah Levine. Introduction to time and frequency metrology. *REVIEW OF SCIENTIFIC INSTRUMENTS*, vol. 70, no. 6, pp. 2567–2596, 1999.

[14] Michael A. Lombardi, Lisa M. Nelson, Andrew N. Novick, and Victor S. Zhang. Time and Frequency Measurements Using the Global Positioning System. *Cal Lab: The International Journal of Metrology*, pp. 26–33, 2001.

[15] CH-2002 Neuchatel Switzerland Oscilloquartz, Rue des Brevards 16. Ocxo 8607: Oven controlled crystal oscillator. Hamilton Technical Services, Beaufort, SC., 2010.

[16] William J. Riley. User manual: Stable32 frequency stability analysis version 1.5.0. Hamilton Technical Services, Beaufort, SC.

[17] William J. Riley. *Handbook of frequency stability analysis.* NIST special publication. Hamilton Technical Services, 2007.

[18] Vojtech Vigner and Jaroslav Roztocil. Device for Time Scale Generation and Event Timestamping. In *IDAACS'2011 — Proceedings of the 6th IEEE International Conference on Intelligent Data Acquisition and Advanced Computing Systems*, pp. 287–290, 2011.

6

FlexRay Modelling and Application*

J. Novák[†], M. Okrouhlý, and J. Sobotka

Czech Technical University, Faculty of Electrical Engineering, Technická 2, 166 27 Prague 6, Czech Republic, [†]jnovak@fel.cvut.cz

Abstract

FlexRay is an incoming standard of automotive distributed system intended for safety critical applications like x-by-wire. The first part of the chapter provides an introduction to the FlexRay standard, presents an analytic model of FlexRay Synchronization Mechanism, validation of this model and offers its usage for measurement of parameters of synchronization mechanism in real FlexRay networks. The second part of the chapter presents a case study of FlexRay application in a distributed implementation of ABS (Anti-lock Braking System) in vehicle. The anti-lock braking system is nowadays a part of almost every vehicle. In order to work properly, the anti-lock braking system needs to have actual information about the rotation of every wheel. Currently, the information about the frequency of wheel rotation is being transmitted via the current loop. This paper describes a method, where the information

*This research is supported by the the Czech Ministry of Education, project name Josef Bozek Research Center of Engine and Automotive Engineering, grant No. 1M0568, Czech Science Foundation (GACR), project name Sensors and intelligent sensor systems, grant No. GD102/09/H082, Czech Science Foundation under the project 102/09/H081 SYNERGY — Mobile Sensoric Systems and Network and Grant Agency of the Czech Technical University in Prague, grant No. SGS12/155/OHK3/2T/13.

Advanced Distributed Measuring Systems — Exhibits of Application, 121–162.

from the rotation sensors of the wheels is transmitted into the control unit through the FlexRay bus. Additional wireless acceleration sensors located on the wheels are used to improve the ABS performance.

Keywords: FlexRay; synchronization mechanism; modeling; ABS.

6.1 Introduction

Today vehicles are equipped with many electronic control systems assisting the driver in daily as well as emergency situations. Electronic control units (ECUs) of these systems are usually interconnected by CAN network. Complexity and number of such control devices has significantly risen during the last decade. Nowadays, the luxury cars contain more than sixty electronic control units interchanging more than two thousand signals. Single CAN bus is not able to satisfy the increasing demands and manufacturers are forced to use the concept of several independent CAN buses interconnected by the gateway [15]. Upcoming x-by-wire systems like steer-by-wire or brake-by-wire require a reliable high-speed communication protocol suitable for such safety critical applications [18], where CAN suffers from non-deterministic MAC [1], limited bandwidth and no physical layer redundancy. For this purpose FlexRay protocol [6] was developed by the FlexRay consortium. The consortium includes (besides others) leaders of the automotive industry like BMW and General Motors, semiconductor industry leaders like Motorola and Philips as well as the vehicle technology leaders like Bosch. The FlexRay is planned to be used as a backbone network in future vehicles. Using this technology for new applications, the amount of wiring and possibly other mechanical and hydraulic components — and thus the costs — can be essentially decreased. Manufacturers of passenger cars play a role of system integrators today. They manufacture only the key components of vehicles (engine, gearbox, chassis ...) and most of other car components are developed and manufactured by third parties according to the vehicle manufacturer specifications. This is especially true for electrical and electronic components. An important task of manufacturer is to test the conformity of delivered third party equipment with the specifications as well as the interoperability among the components. Such tests are running during most vehicle development phases. Appropriate methods and instruments are necessary for objective and repeatable results of these tests, especially if a new technology (e.g., the FlexRay) is being incorporated [9]. For the CAN communication, vehicle manufacturers perform hundreds of tests focused on the correct behavior of a single ECU on the CAN bus and thousands functional and integration tests focused on the functionality of particular vehicle subsystems, including the ECUs interaction via communication infrastructure. The CAN tests are performed at physical, data link and application protocol layers and they are proved by the long time experience with this standard [16]. The situation is rather different for the FlexRay, especially for data link layer tests. FlexRay data link layer protocol is much more complex than that of the CAN and many parameters have to be set by the ECU firmware [11]. Incorrect setting of these parameters is sometimes invisible at the first view, but it can lead to the decreased system performance or even to the ECU or complete cluster failure under the specific conditions. The test methods and appropriate instruments are thus necessary to allow independent testing of the data link layer parameters programmed into FlexRay controllers in particular ECUs. The first step to reach this goal is to develop models of particular algorithms used at the FlexRay data link layer, validate them and finally use them for development of the required measurement methods.

The aim of the second part of this chapter is to present an application of the new communication protocol in ABS by replacing the existing connection between the sensors and ECU and to improve ABS control algorithm based on acceleration obtained from wireless acceleration sensor. The automotive anti-lock braking system (ABS) is an active safety technology which is currently the most popular and effective one. ABS can significantly improve the direction stability and steering control ability while braking and also reduce the braking distance. Due to its great contribution to the automotive safety, further research of this technology is desirable. The main disadvantage of the currently used anti-lock braking systems appears during braking on a disrupted roadway. When the vehicle is braking on this roadway type, the wheels often lose contact with the road surface. In this case the wheel is blocked due to pressure in the hydraulic system and the situation is classified by the ABS system as a slip, so the ABS system is activated. Now, the activated anti-lock braking system opens the output valve, so that the pressure in the hydraulic system decreases and the wheel is thus unblocked. Due to wheel unblocking, the brake distance is significantly increased. In case of known information from the acceleration sensor located on the wheel, it can be assumed that the wheel does not rotate due to decreased clamping force and can be locked before it touches the road surface again. Hereby the brake distance is decreased. The algorithm is evaluated on a developed ABS test platform.

6.2 FlexRay standard description

In this section the FlexRay standard [6] is briefly described, especially the synchronization mechanism related to the first part of below presented work. The standard was established by FlexRay consortium in 1999. FlexRay is primarily intended as a communication standard for automotive x-by-wire applications, but today it is mostly used for active chassis control systems. The block diagram of a FlexRay node is shown in Figure 6.1. The block diagram of a node can be divided into 3 parts corresponding to the three protocol layers of ISO/OSI communication model.

- Physical layer is implemented by a FlexRay transceiver with optional element called Bus Guardian [21], which should prevent an erroneous access to the bus outside of the dedicated time slot, assigned to the node (solves so called bubbling idiot error). There are up to two FlexRay transceivers in the node, as FlexRay supports two independent redundant channels (A and B).

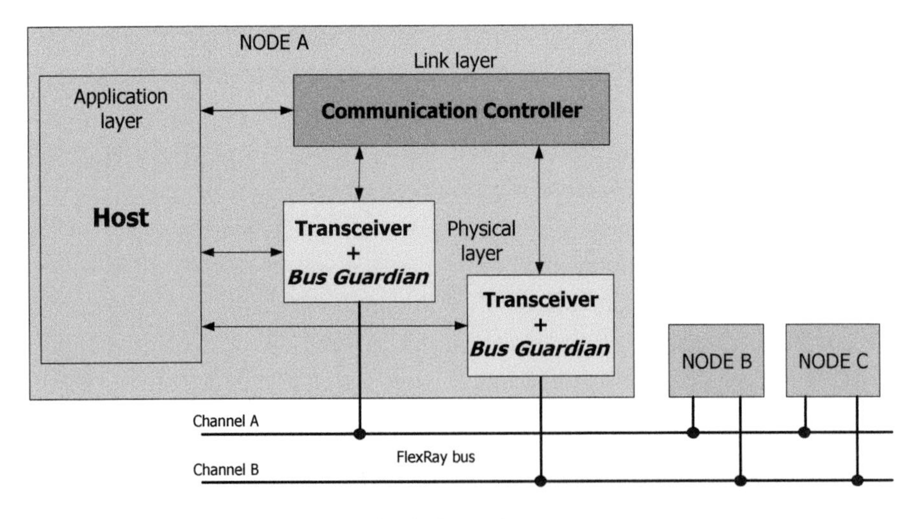

Fig. 6.1 FlexRay node structure.

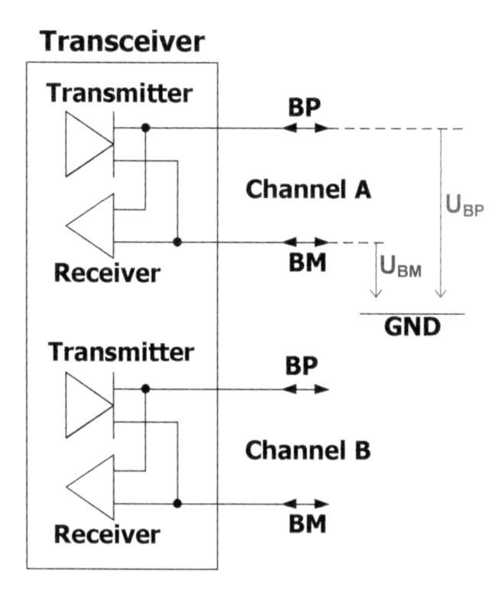

Fig. 6.2 FlexRay physical interface.

- Data link layer is represented with the communication controller implementing FlexRay communication protocol described by standard.
- Application layer protocols are implemented in the host. The host provides two main functions. It configures the communication controller with required setting associated with communication parameters (physical and data link layer parameters) first. Second, the application software is started, which functionality and protocols are out of the range of the FlexRay standard.

Physical layer is quite similar to CAN or RS485 and uses differential signaling (see Figure 6.3). Nodes are usually connected by twisted pair cable. Passive linear network, passive star, active star (and their combination) network topologies are supported. Two channels structure is supported - particular node can be connected to channel A, channel B or both channels. Maximum communication speed defined by the standard is 10 Mbps.

Data link layer communication is based on TDMA (time division multiple access) method and it is running by means of communication cycles. Communication cycle is divided into four parts called segments. Static segment (mandatory), dynamic segment (optional), symbol window segment (optional) and idle segment (mandatory, see Figure 6.4).

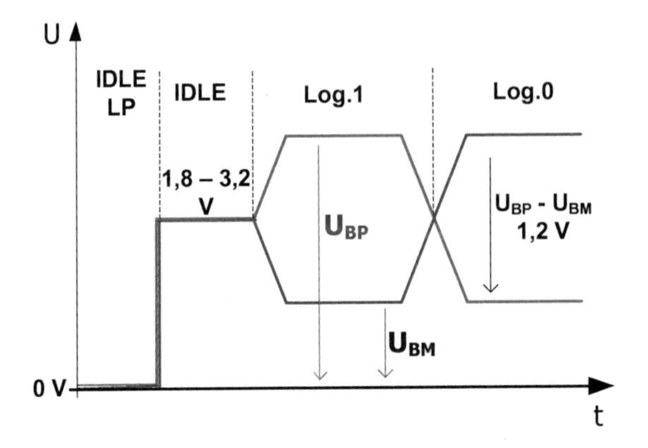

Fig. 6.3 FlexRay physical signaling.

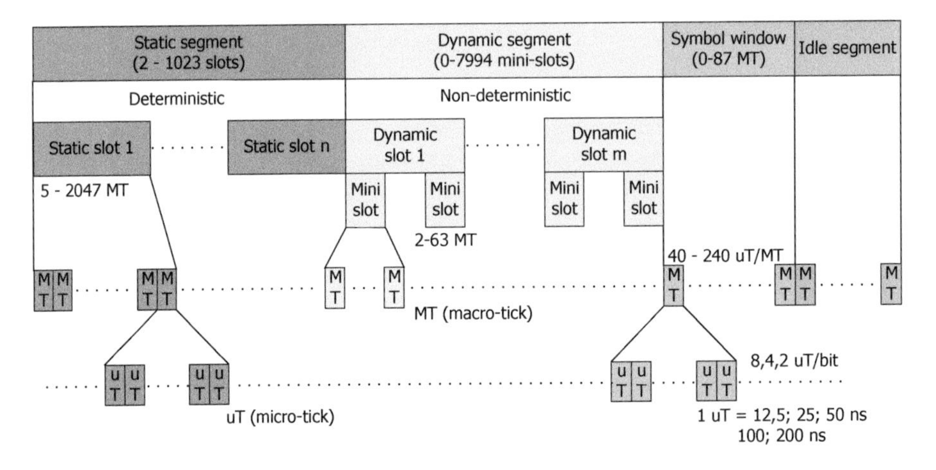

Fig. 6.4 FlexRay communication cycle.

The mandatory static segment comprises a fixed number of equally sized TDMA static slots. The segment is designed for communication where predictable frame delay is required. In each static slot a frame (with a unique identifier) can be transmitted by a node. Once the static slot is assigned to a specific node, only this node is allowed to transmit frame within this slot and thus the protocol prevents collisions. Communication within static segment is deterministic and can be used for hard real-time applications [22]. The optional dynamic segment employs the flexible TDMA approach. It is intended for the non-critical, sporadic event-driven messages with varying length. Dynamic slots are built-up from variable number of mini-slots. The length of a dynamic slot is flexibly adapted to the length of data in a particular data frame. This is the main difference compared to the static slots. This results in fact that the time where particular dynamic slot begins is not exactly known, as it depends on volume of communication in previous dynamic slots. In addition, the length of dynamic segment is constant. It means the communication, situated in later dynamic slots in the dynamic segment, can be postponed to dynamic segment in the next communication cycle. The postponing can be performed more times until the volume of communication in previous dynamic slots within dynamic segment decreases enough. The optional symbol window segment is designed for transmission of so-called symbols - arbitrary sequences of bits that do not respect the frame format. The mandatory idle segment is used to isolate two successive communication cycles (see Figure 6.4). There is physical layer IDLE state (see Figure 6.3) on the bus.

The communication timing is based on two time units:

- **Micro-tick** (μT)
  Micro-tick is the smallest indivisible time unit in the FlexRay protocol. The unit is directly derived from the time base in the node. One μT is usually equal to the period of the clock signal. For example, if frequency of clock signal is 40 MHz, 1 μT corresponds to 25 ns.
- **Macro-tick** (MT)
  Macro-tick is a longer time unit consisting of integer number of μT. One MT is a global time unit in network. The duration of communication cycle as well as all its parts is defined in integer number of MT. The nominal numbers must be the same in each node of the network. On the other hand, the μT is a local time unit (derived from local oscillator) and can be of different length in particular nodes. Consequently, the number of μT falling into single MT is different within nodes.

Data among nodes are exchanged by means of data link frames. The frame consists of three parts — the header segment, the payload segment, and the CRC segment (see Figure 6.5).

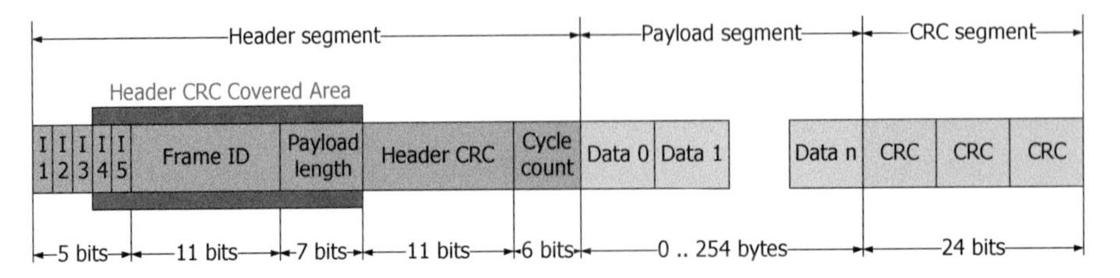

Fig. 6.5 FlexRay frame format.

The frame ID is assigned the slot in which the frame should be transmitted. A frame ID may be used no more than once on each channel within a communication cycle. The payload length field is used to indicate the size of the payload segment. The header CRC contains a cyclic redundancy check code that is computed over the synchronization frame indicator (I4), the start-up frame indicator (I5), the frame ID and the payload length. The cycle count indicates the transmitting node view of the value of the communication cycle counter at the time of frame transmission. The payload segment contains up to 254 bytes of data. The frame CRC segment contains a cyclic redundancy check code computed over the all fields of the header segment and the payload segment.

6.2.1 Synchronization Mechanism

Each FlexRay node has its own time base driven by a quartz-crystal oscillator [2]. Due to some influences such as crystal ageing or temperature variation the instantaneous frequencies of oscillators in particular network nodes are different. Without unceasing corrections in all nodes the network would not be able to work at all because the slot boundary violations occur. The synchronization mechanism, incorporated in the data link layer protocol, is responsible for these corrections. From the point of view of the role in the synchronization mechanism FlexRay network consists of two types of node. First type sends and receives synchronization frames, the second type only receives synchronization frames. Synchronization frame is a standard frame with synchronization frame attribute (I4 bit in a header segment) set. The main task for clock synchronization mechanism in a FlexRay network is to bring the time basis to have approximately the same view of time. Each FlexRay network must contain at least two synchronization nodes transmitting synchronization

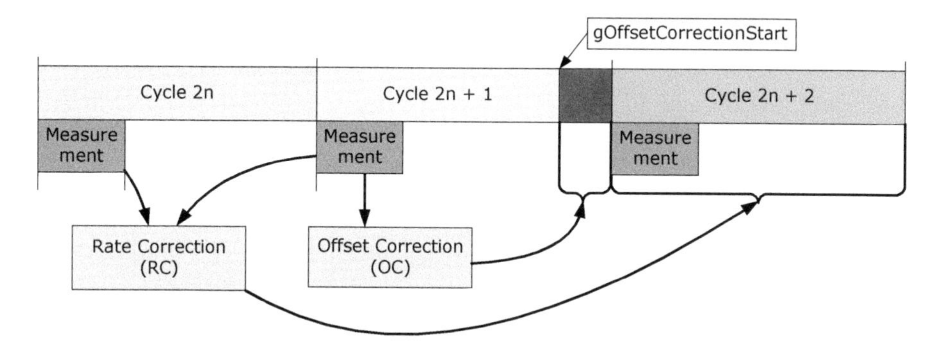

Fig. 6.6 Principle of synchronization mechanism in FlexRay.

frames. The correction mechanism in every node is based on measuring of time deviations between the expected and actual time of synchronization frames arrival from the other synchronization nodes in appropriate time slots in static segment. The principle is shown in Figure 6.6.

The measurement of deviations takes place in each communication cycle in every node. FlexRay synchronization mechanism uses two types of corrections — rate and offset corrections; they are both computed using these deviations.

First, some terms and variables used in this section should be explained:

OC Offset Correction

RC Rate Correction

fix(x) Mathematical operation truncating a decimal part of real number x to zero

vMicrotick [μT] Counter of micro-ticks. This variable shows actual number of μT within a communication cycle.

vMacrotick [MT] Counter of macro-ticks. This variable shows actual number of MT within a communication cycle.

vCycleCounter [$-$] Counter of communication cycles. This variable shows actual communication cycle number. The range is from 0 to 63.

gOffsetCorrectionStart [MT] Number of MT from start of the communication cycle where the application of OC starts.

pMicroPerCycle [μT] Nominal number of μT for one communication cycle.

zMicroPerPeriod [μT] Modified number of μT of pMicroPerCycle. This modification is made by correction values RC and OC.

gMacroPerCycle [MT] Number of MT for one communication cycle.
zMacroPerPeriod [MT] Number of MT within the part of communication cycle where particular micro-tick distribution is made. When RC is distributed only, zMacroPerPeriod equals to gMacroPerCycle. When RC together with OC are distributed, zMacroPerPeriod equals to difference gMacroPerCycle — gOffsetCorrectionStart.
pOffsetCorrectionOut [μT] Maximum permissible offset correction
pRateCorrectionOut [μT] Maximum permissible rate correction
pClusterDriftDamping [μT] Damping factor is used for the rate correction. The computed RC parameter is always changed by pClusterDriftDamping factor towards zero. If absolute value of computed RC parameter is lower than pClusterDriftDamping, resulting RC value is zero. This mechanism serves for avoiding the instability in rate correction mechanism.

Generally, the RC value is given as a difference between two measurements in two successive communication cycles (the first of them is even and the second is odd). RC corresponds to rate difference between the local time base and time basis in the remaining network nodes. The OC value is determined from measurement in each communication cycle and represents a remaining time offset between the local time base and time basis in the rest of the network. Based on the correction values, the length of communication cycle is lengthened or shortened it depends on whether the local time base is slower or faster than the time bases in other network nodes. This ensures the nodes have a similar view of global time, even if there are some frequency differences among the nodes oscillators. Now let us describe the synchronization mechanism step by step in details. The mechanism is divided into two main processes.

Clock Synchronization Process

Clock synchronization process measures time deviations between supposed and actual arrival time of synchronization frames coming from other nodes [10]. These deviations are used to compute the correction values RC and OC. The flowchart describing this process is shown in Figure 6.7.

Measurement of deviations takes place at the beginning of each communication cycle, more precisely within the static segment where synchronization frames are received in particular static slots. Each deviation is determined in μT as a difference between actual and expected frame arrival time. After

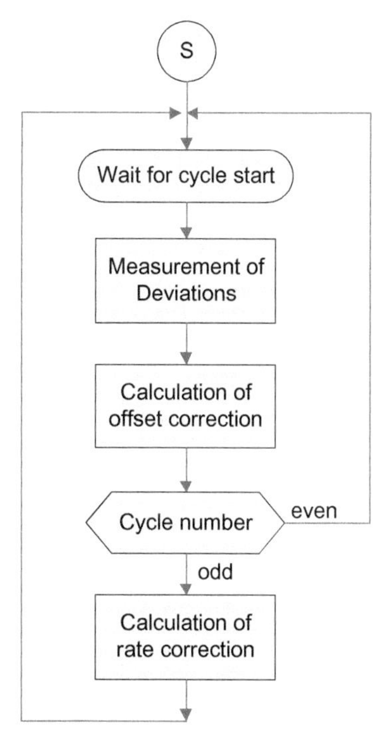

Fig. 6.7 Block diagram of clock synchronization process.

receiving all synchronization frames, each node has a table filled with the measured deviation values. On the basis of measured deviation table, each node calculates offset correction value in each communication cycle. The rate correction value is computed in odd communication cycles only. In even cycles, the rate correction value is taken from the previous odd cycle.

The detailed algorithm for computation of offset correction value is in Figure 6.8.

In the first step, the minimum value for each node is chosen from values received at channels A and B and the table of deviations called DevList, containing N deviations got from N synchronization nodes, is prepared. The offset correction value is evaluated in the second step from the DevList table using fault-tolerant midpoint algorithm **midterm**. Its principle is shown in Figure 6.9.

The deviations are ordered by size. The k smallest and the k biggest values are discarded, where k depends on the number N of measured deviations

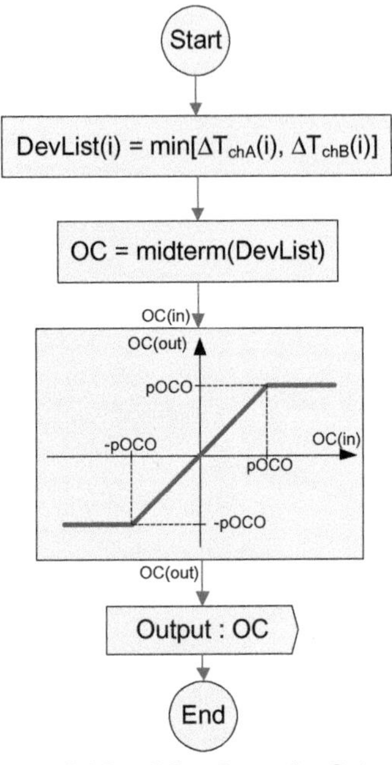

Fig. 6.8 Offset correction value evaluation.

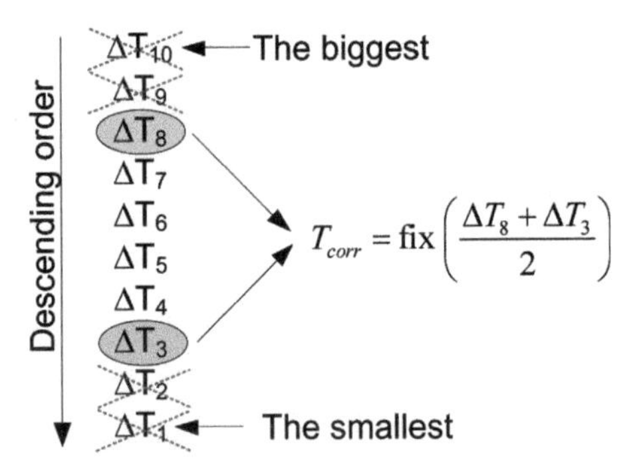

Fig. 6.9 Fault tolerant midpoint algorithm.

($k = 0$ for $N < 3, k = 1$ for $N \geq 3$ and $N < 8, k = 2$ for $N \geq 8$). The smallest and the biggest values are chosen from this reduced table, their average value is computed and decimal part is truncated.

Finally, the value is limited by factor pOffsetCorrectionOut. Now the offset correction value is found.

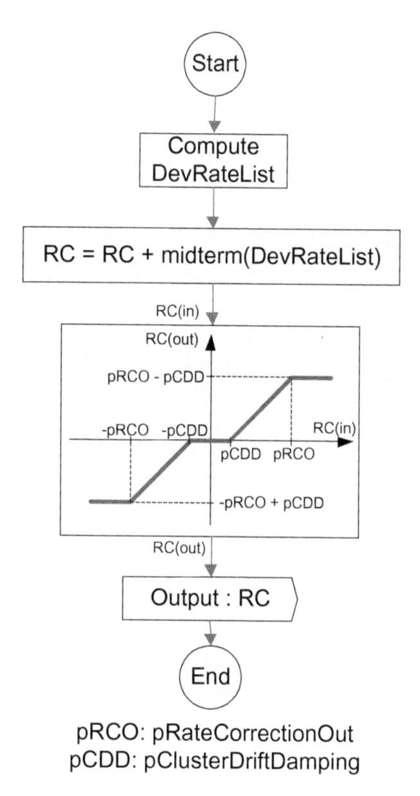

Fig. 6.10 Rate correction value evaluation.

The detailed algorithm for computation of rate correction value is in Figure 6.10.

In the first step the DevRateList table is prepared. The difference between deviations in odd and even communication cycles is computed for channel A and channel B and their average value is evaluated for each synchronization node. The DevRateList table then contains N deviation differences for N synchronization nodes. In the second step the rate correction value is evaluated from this table by means of fault-tolerant midpoint algorithm (see Figure 6.9). Finally, the value is damped by factor pClusterDriftDamping and limited by factor pRateCorrectionOut in the third step.

Macro-tick Generation Process

Macro-tick generation process is responsible for application of computed rate and offset corrections within the communication cycles. The communication cycle can be lengthened or shortened by this process. The way in which the correction values are applied is described in this paragraph. The rate correction is a number of μT computed on the basis of deviations measured in an even and odd successive communication cycles (i.e., cycles number 2n and 2n + 1). The value is applied in subsequent even (2n + 2) and odd (2n + 3) communication cycles. Application means that the rate correction value is evenly distributed over the whole communication cycle. This distribution is performed in such a way that nominal number of μT per MT in evenly chosen MT is modified by 1 μT. The offset correction value is a number of μT computed on the base of measured deviations in each odd communication cycle. The value is applied at the end of the same odd cycle in the same way as rate correction value is applied. However, instead of even distribution over the cycle, the offset correction value is distributed only in an interval from gOffsetCorrectionStart to the end of the communication cycle (see Figure 6.6). Both the rate and offset correction values can be positive or negative, thus the cycle can be lengthened or shortened.

6.2.2 Start-up Mechanism

The necessity of the start-up procedure should be mentioned as a minor disadvantage of the FlexRay protocol [13]. As the start-up mechanism is not a subject of this chapter, it is described briefly. To form a cluster at least two nodes are necessary. The initialization of the start-up process is called cold-start. The nodes that are able to start this process are called cold-start nodes; the other non-cold-start nodes. 2 up to 15 cold-start nodes are possible. The cold-start node starts by transmission of a CAS symbol. Only the cold-start node transmitting the CAS (leading cold-start node) may transmit frames within the first four communication cycles after the CAS. Consequently, the remaining cold-start nodes are joined into the communication cycle and afterwards also the rest of the nodes (non-cold-start). In case there is no activity on a bus (before the start-up procedure), one node can wake up a cluster. The minimum prerequisite for a cluster wakeup is that the receivers of all bus drivers are supplied with power. A bus driver has the ability to wake up the other components of its node when it receives a wakeup pattern (WUS) on its channel.

6.3 Model of FlexRay Synchronization Mechanism

Fault-free operation of FlexRay Synchronization Mechanism is crucial for deployment FlexRay communication standard in safety critical applications like x-by-wire systems. Investigation of the synchronization mechanism reliability will be quite easier if we have the model that allows to study the behaviour of synchronization mechanism in different situations. Such a model can be also useful for development of algorithms for evaluation of parameters related to synchronization mechanism.

There are two ways how to make the model of synchronization mechanism - behavioural model and analytic model. The behavioural model can be implemented as a set of operations described in the FlexRay standard. The simulation in such a model is very time consuming, especially when a large network with slightly different clock frequencies in particular nodes is modelled. This is why we decided to design an analytic model. Instead of performing so many steps, the state of the system, in the desired part of communication cycle, is computed directly using derived mathematical formulas. In fact, the simulation in this model is much faster. In a simplified way it is possible to say that the number of steps is reduced to 1 per one communication cycle without reducing the resolution in frequency. The input for the model is the set of network parameters. First the particular deviations in each node are calculated, and then the rate correction (RC) and offset correction (OC) values in each communication cycle are evaluated. The results are evaluated repeatedly for each communication cycle.

6.3.1 Model Derivation

In this section, step by step derivation of analytic model will be presented. First of all, define some terms and shortcuts used in the mathematical model is necessary:

AMT

is number of macro-ticks between the start of cycle and the start of an action in the cycle. The action can represent any event on the bus (for example an arrival of synchronization frame or the end of communication cycle).

MPCMT
is the number of Macro-ticks per one communication cycle.

OCSMT
Offset Correction Start is number of macro-ticks from start of odd communication cycle up to the point where the application of OC starts.

$\mu PC_{nom}\mu$T
is nominal number of μT per one communication cycle.

$\mu PM_{nom}\mu$T/MT
is nominal number of μT per one MT. Following term is valid.

$$\mu PM_{nom} = \frac{\mu PC_{nom}}{MPC}$$

$RC^i\mu$T
is computed value of rate correction in i^{th} communication cycle.

$OC^i\mu$T
is computed value of offset correction in i^{th} communication cycle.

$\mu PA_x^i(A)\mu$T
is the number of micro-ticks between the start of cycle and start of an action A in i^{th} communication cycle in node x. The value is modified by correction parameters RC^i and OC^i.

$\Delta\mu PA_{xy\to z}^i(A)\mu$T
expresses a deviation between time of detection of action A in node y and in node x in perspective of reference node z in i^{th} communication cycle.

$\Delta\mu PA_{xy\to z}^i(EOC)\mu$T
is $\Delta\mu PA_{xy\to z}^i(A)$, where $A^i = EOC^{i-1}$. The EOC^{i-1} is action representing end of previous cycle (cycle $i-1$).

$\mu PA_{add_RC}^i(OCS)\mu$T
corresponds to relevant proportion of RC^i parameter distributed in interval from the point — start of cycle to the point — start of offset correction in i^{th} communication cycle.

$\mu PA_{add_RC}^i(A - OCS)\mu$T
is relevant proportion of RC^i parameter that falls in interval from the point — start of offset correction (OCS) to the point — start of an action A in i^{th} communication cycle.

$\mu PA^{i}_{add\_OC}(A - OCS)\mu\mathrm{T}$
is relevant proportion of OC^i parameter that falls in interval from the point — start of offset correction (OCS) to the point — start of an action A in i^{th} communication cycle.

$\mu PA^{i}_{add}(A)\mu\mathrm{T}$
is relevant proportion of both corrections applied to the point — start of an action A in i^{th} communication cycle.

From section 6.2.1 is obvious, that for determination of both correction values it is necessary to know difference between expected and real synchronization frame arrival time. Reference node z was defined for the derivation. Parameters of node z are equal to nominal. Time difference between occurrence of action A in nodes x and y from the reference node point of view in i^{th} communication cycle is defined by following term:

$$\Delta\mu PA^{i}_{xy\to z}(A) = k_{y\to z}\cdot\mu PA^{i}_{y}(A) - k_{x\to z}\cdot\mu PA^{i}_{x}(A) + \Delta\mu PA^{i-1}_{xy\to z}(EOC), \quad (6.1)$$

where:
$k_{y\to z} = f_z/f_y-, k_{x\to z} = f_z/f_x-,$
f_z is reference node local oscillator frequency [Hz],
f_y is node y local oscillator frequency [Hz],
f_x is node x local oscillator frequency [Hz].

Initial condition:

$$\Delta\mu PA^{-1}_{xy\to z}(EOC) = 0 \quad (6.2)$$

Correction applied in individual nodes is described by following term:

$$\mu PA^{i}(A) = \mu PA_{nom}(A) + \mu PA^{i}_{add}(A) \quad (6.3)$$

The first summand is equal to:

$$\mu PA_{nom}(A) = A\cdot\mu PA_{nom} \quad (6.4)$$

Expression of second summand in formula 6.3 depends on type of applied correction. If only rate correction is applied — action occurs in even communication cycle or in odd communication cycle before offset correction

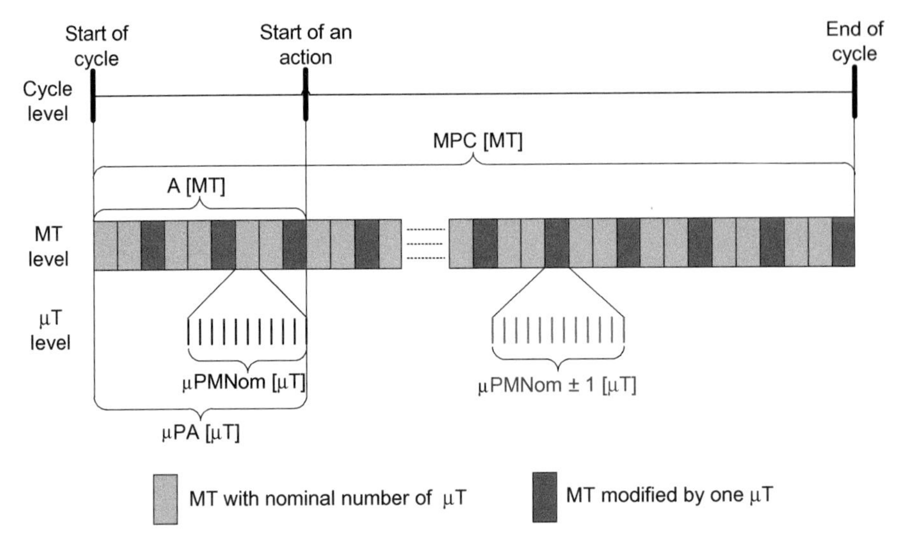

Fig. 6.11 Distribution of RC value in an even CC.

start point (see Figure 6.11), the term 6.5 is valid. This situation will be referred as variant I.

$$\mu PA^{i}_{add}(A) = \frac{A}{MPC} \cdot RC^{i} \tag{6.5}$$

$$\mu PA^{i}(A) = A \cdot \mu PA_{nom} + \frac{A}{MPC} \cdot RC^{i} \tag{6.6}$$

In second case (referred as variant II, depicted in Figure 6.12), when monitored action occurs in odd communication cycle after offset correction start point ($A > OCS$). The situation is expressed by Equation 6.7.

$$\begin{aligned} \mu PA^{i}_{add}(A) = \mu PA^{i}_{add\_RC}(OCS) + \mu PA^{i}_{add\_RC}(A - OCS) \\ + \mu PA^{i}_{add\_OC}(A - OCS) \end{aligned} \tag{6.7}$$

Individual members of the sum are equal to:

$$\mu PA^{i}_{add\_RC}(OCS) = \frac{OCS}{MPC} \cdot RC^{i} \tag{6.8}$$

$$\mu PA^{i}_{add\_RC}(A - OCS) = \frac{A - OCS}{MPC - OCS} \cdot \left(RC^{i} - \frac{OCS}{MPC} \cdot RC^{i} \right) \tag{6.9}$$

$$\mu PA^{i}_{add\_OC}(A - OCS) = \frac{A - OCS}{MPC - OCS} \cdot OC^{i} \tag{6.10}$$

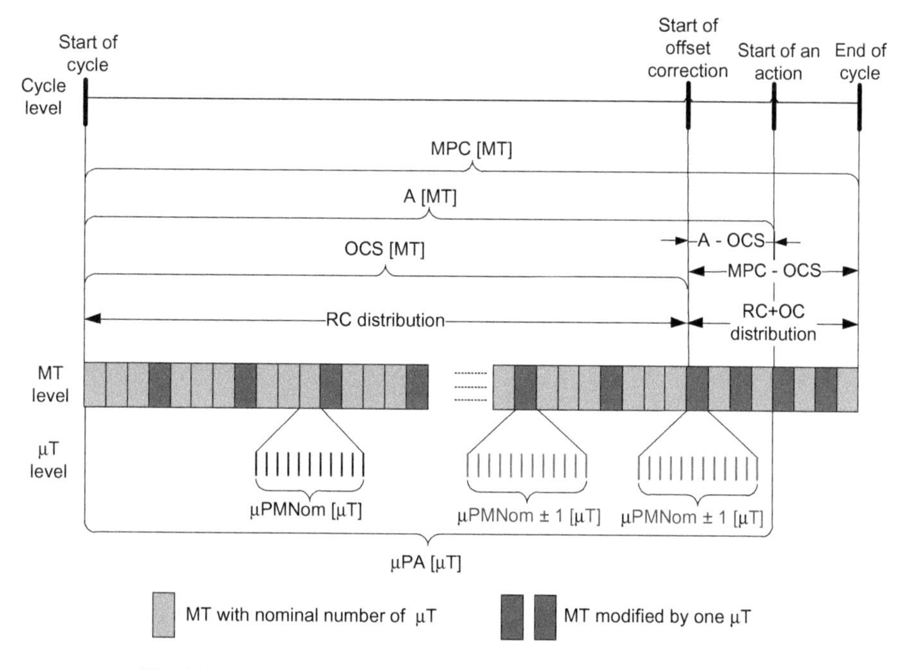

Fig. 6.12 Distribution of RC and OC correction parameters in an odd CC.

Final formula for variant I is defined by Equation 6.11, Equation 6.12 respects actual hardware time resolution of 1 μT.

$$\begin{aligned}\Delta\mu PA^{i}_{xy\to z}(A) = {} & \frac{f_z}{f_y}\cdot\left(A\cdot\mu PM_{nom}+\frac{A}{MPC}\cdot RC^{i}_{y}\right)\\ & -\frac{f_z}{f_x}\cdot\left(A\cdot\mu PM_{nom}+\frac{A}{MPC}\cdot RC^{i}_{x}\right)\\ & +\Delta\mu PA^{i-1}_{xy\to z}(EOC)\end{aligned} \tag{6.11}$$

$$\Delta\mu PA^{i}_{xy\to z}(A) = \mathrm{fix}\left[\begin{array}{l}\frac{f_z}{f_y}\cdot\left(A\cdot\mu PM_{nom}+\mathrm{fix}\left(\frac{A}{MPC}\cdot RC^{i}_{y}\right)\right)\\ \quad -\frac{f_z}{f_x}\cdot\left(A\cdot\mu PM_{nom}+\mathrm{fix}\left(\frac{A}{MPC}\cdot RC^{i}_{x}\right)\right)\\ \quad +\Delta\mu PA^{i-1}_{xy\to z}(EOC)\end{array}\right] \tag{6.12}$$

Final formula for variant II is defined by Equation 6.13, Equation 6.14 respects again actual hardware time resolution of 1 μT.

$$\Delta\mu PA^{i}_{xy\to z}(A) = \frac{f_z}{f_y} \cdot \begin{bmatrix} A \cdot \mu PM_{nom} + \frac{OCS}{MPC} \cdot RC^{i}_{y} \\ + \frac{A - OCS}{MPC - OCS} \cdot \left(RC^{i}_{y} - \frac{OCS}{MPC} \cdot RC^{i}_{y}\right) \\ + \frac{A - OCS}{MPC - OCS} \cdot OC^{i}_{y} \end{bmatrix} - \frac{f_z}{f_x} \cdot \begin{bmatrix} A \cdot \mu PM_{nom} + \frac{OCS}{MPC} \cdot RC^{i}_{x} \\ + \frac{A - OCS}{MPC - OCS} \cdot \left(RC^{i}_{x} - \frac{OCS}{MPC} \cdot RC^{i}_{x}\right) \\ + \frac{A - OCS}{MPC - OCS} \cdot OC^{i}_{x} \end{bmatrix} + \Delta\mu PA^{i-1}_{xy\to z}(EOC) \tag{6.13}$$

$$\Delta\mu PA^{i}_{xy\to z}(A) = \mathrm{fix}\begin{bmatrix} \frac{f_z}{f_y} \cdot \begin{bmatrix} A \cdot \mu PM_{nom} + \mathrm{fix}\left(\frac{OCS}{MPC} \cdot RC^{i}_{y}\right) \\ + \mathrm{fix}\left(\frac{A - OCS}{MPC - OCS} \cdot \left(RC^{i}_{y} - \mathrm{fix}\left(\frac{OCS}{MPC} \cdot RC^{i}_{y}\right) + OC^{i}_{y}\right)\right) \end{bmatrix} \\ - \frac{f_z}{f_x} \cdot \begin{bmatrix} A \cdot \mu PM_{nom} + \mathrm{fix}\left(\frac{OCS}{MPC} \cdot RC^{i}_{x}\right) \\ + \mathrm{fix}\left(\frac{A - OCS}{MPC - OCS} \cdot \left(RC^{i}_{x} - \mathrm{fix}\left(\frac{OCS}{MPC} \cdot RC^{i}_{x}\right) + OC^{i}_{x}\right)\right) \end{bmatrix} \\ + \Delta\mu PA^{i-1}_{xy\to z}(EOC) \end{bmatrix} \tag{6.14}$$

Derived equations are based on simplifying assumptions, that frequency of local oscillators f_x and f_y are constant on interval from communication cycle to the point where action A occurs (for frequency of oscillator f_z this assumption is valid). Real model of course reflects the time dependence of local oscillator frequency and values f_x and f_z are replaced by mean value by Equation 6.15.

$$\bar{f} = \frac{1}{T_A - T_0} \cdot \int_{T_0}^{T_A} f(t) \cdot dt \tag{6.15}$$

T_0 is time of start of communication cycle,
T_A is time of occurrence of action A.

Derived Equations 6.12 and 6.14 are used to build the tables of time differences between expected and actual arrival times of synchronization frames in particular nodes. Offset and rate correction values are evaluated in our model according to algorithm described in 6.2.1.

6.3.2 Model Implementation

Derived analytic model can be implemented in any high-level programming language. We are using implementation in plain C language. Final application allows simulation in predefined number of communication cycles. Parameters of each node are represented by a C structure. Simulation of behaviour of FlexRay network consisting from nodes with different parameters is thus possible. Parameters related to the synchronization mechanism (described in previous section) are used as an input and output is available as a set of text files — one for each simulated node. The file contains values of rate correction, offset correction and table of deviations (DevList) in all simulation step. Produced values can be used for synchronization mechanism behaviour investigation. Example of produced values with comparison to measurements from actual network is shown in next section about model validation.

6.3.3 Model Validation

Plausibility of the model was investigated by comparing results of tests of synchronization mechanism obtained by simulation and by measurement on real network (actual values of rate and offset correction in particular communication cycles are stored locally within the nodes and downloaded after the experiment ends). Model validation was performed in the frame of dissertation thesis [14]. Block diagram of network with adjustable clock sources is

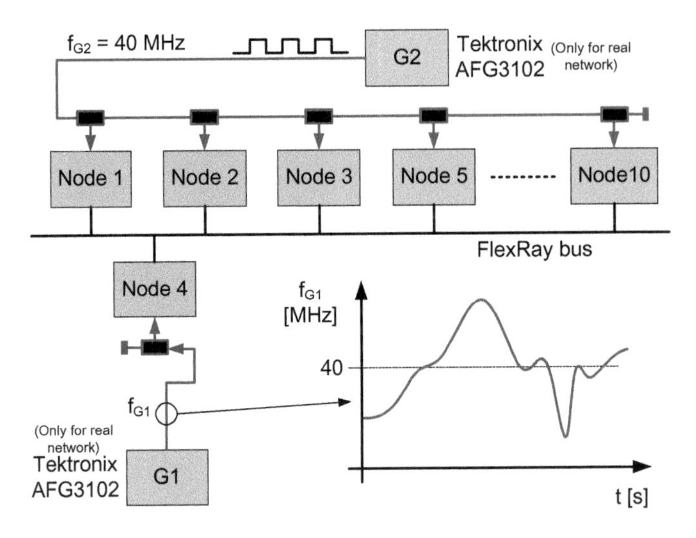

Fig. 6.13 Block diagram of testing network.

depicted in Figure 6.13. Real FlexRay network consists of ten nodes in active star topology. As a source of the external clock the programmable generator was used. Generator G1 is used as clock source for node No. 4. Clock source for remaining nodes is generator G2. According to generator specification, the accuracy of output frequency and linearity error, when the frequency is linearly swept, can cause differences between measured and theoretical results. However, these differences are not caused by imperfection of model. This must be taken into account when plausibility of model is analysed. First static test was performed.

Static Test

In this test, the clock frequency of all nodes was stable (stability is determined by the quality of generators) and clock frequency of node No. 4 was changed with step 200 Hz.

Rate correction value:

Table 6.1 shows that results got from the real system are the same as from the model in a wide range of frequencies. The necessary quantum of frequency for changing of the RC value is 200 Hz and the maximal error of output frequency from the generator is ±40 Hz. Thus, the measured RC value can be influenced

Table 6.1. Measured and simulated rate correction values in node No. 4.

f [MHz]	RC [μT] (simulation)	RC [μT] (measurement)	f [MHz]	RC [μT] (simulation)	RC [μT] (measurement)
39, 9	−499	−499	40, 01	49	49
39, 91	−449	−449	40, 02	99	99
39, 92	−399	−399	40, 03	149	149
39, 93	−349	−349	40, 04	199	199
39, 94	−299	−299	40, 05	249	249
39, 95	−249	−249	40, 06	299	299
39, 96	−199	−199	40, 07	349	349
39, 97	−149	−149	40, 08	399	399
39, 98	−99	−99	40, 09	449	449
39, 99	−49	−49	40, 1	499	499
40	0	0			

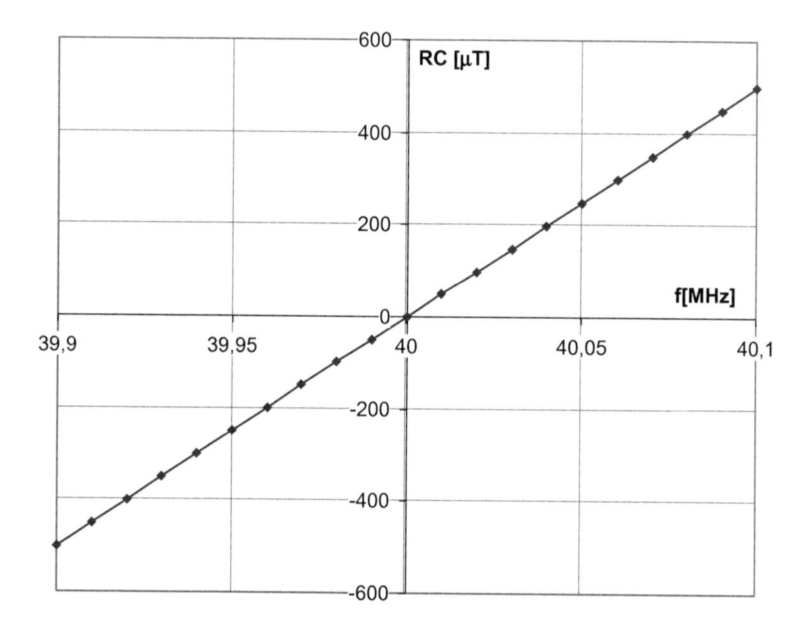

Fig. 6.14 Dependency of simulated and measured RC values on frequency in node No. 4.

by the generator in range $\pm 1\mu$T, when the preset frequency is placed on the boundary of two frequency quanta.

Offset correction value:

The results of the test for offset correction value are depicted in Figure 6.15. Presented OC values are obtained by simulation in model (red bars) as well as by measurement in real system (blue bars). The values are defined in odd

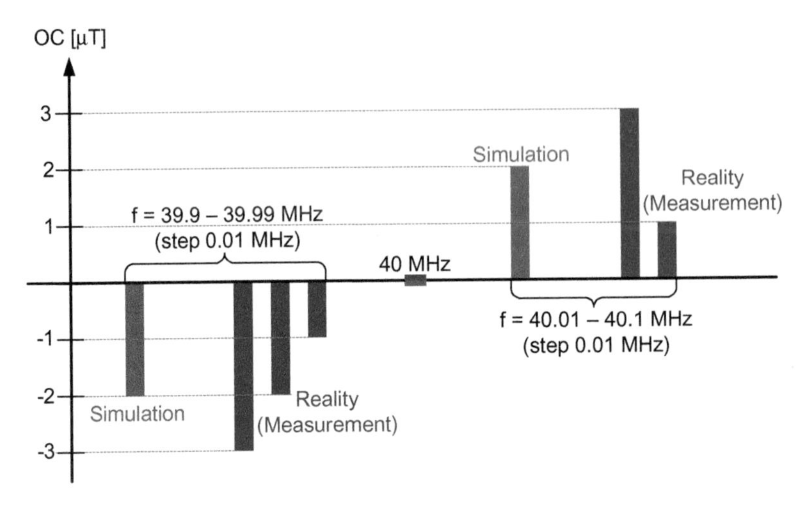

Fig. 6.15 Measured and simulated OC values in node No. 4.

communication cycle, where they are also applied. According to sign of offset correction values the results can be divided into three groups.

The first group contains negative values and is observed for frequency interval $< 39.9; 40)$ MHz. The measured OC values reach $-3, -2, -1\mu$T in three successive odd communication cycles and it repeats periodically. It is clear that the average value is -2μT. Thus, in this group the behaviour of offset correction mechanism in real system and in model is the same (the μT difference between model and real system is caused by quantization of time — in model everything runs synchronously, in real system this state is unreachable). The second group covers OC values equal to zero. The measured OC value and the simulated OC value are equal to zero in each odd communication cycle. Thus, in this group the behaviour of offset correction mechanism in real system and in model is the same. The third group consists of positive values and is defined for frequency interval $(40; 40.1 >$ MHz. The measured OC values reach $+1, +3\mu$T in two successive communication cycle and it repeats periodically. It is clear that the average value is $+2\mu$T. Thus, in this group the behaviour of offset correction mechanism in real system and in model is the same.

Dynamic Test

In case of dynamic test, the frequency f_{G1} was swept. During the test, the clock frequency in node No. 4 is swept by linear shape consisting of two

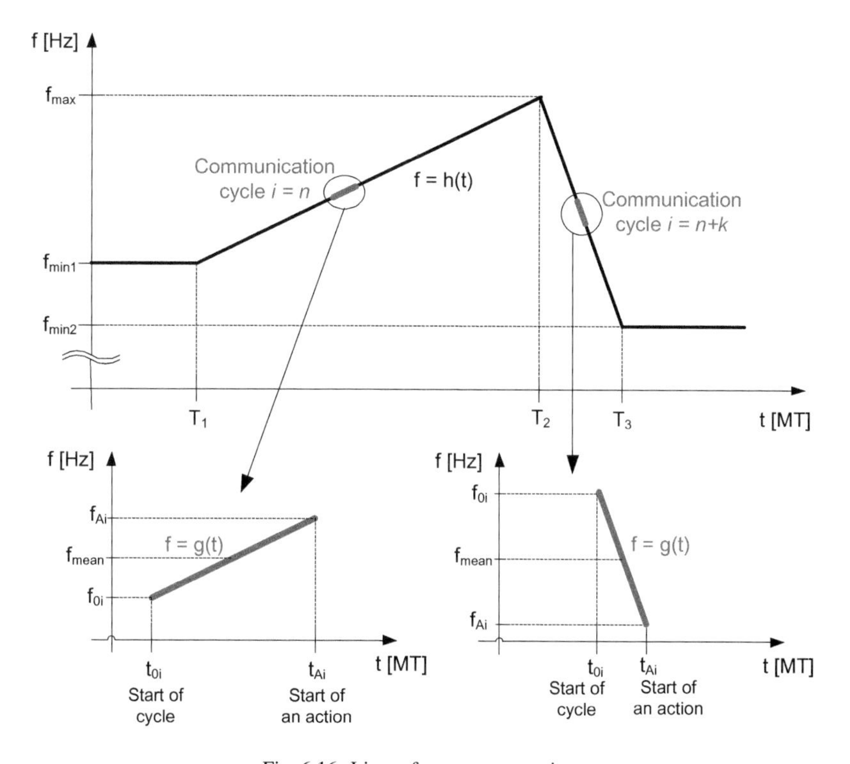

Fig. 6.16 Linear frequency sweeping.

parts increasing and descending with different steepness. Principle of linear sweeping is depicted in Figure 6.16. The test was made on the real system as well as on the model. The results are depicted in Figures 6.17–6.22. Each graph contains two dependencies. The first of them is measured or simulated correction value (OC, RC, ΔOC, ΔRC) dependence on cycle number (CN) — blue colour. The second of them is dependency of frequency (f) on cycle number (CN) — green colour. Each depicted dependency was applied in node No. 4.

As mentioned above the differences between measured and simulated values of corrections are caused by by quantization of time — in model everything runs synchronously, in real system this state is unreachable.

6.3.4 Model Application

Presented model can be used for many different purposes. The common task in automotive industry is to test correct function of Electronic Control Units

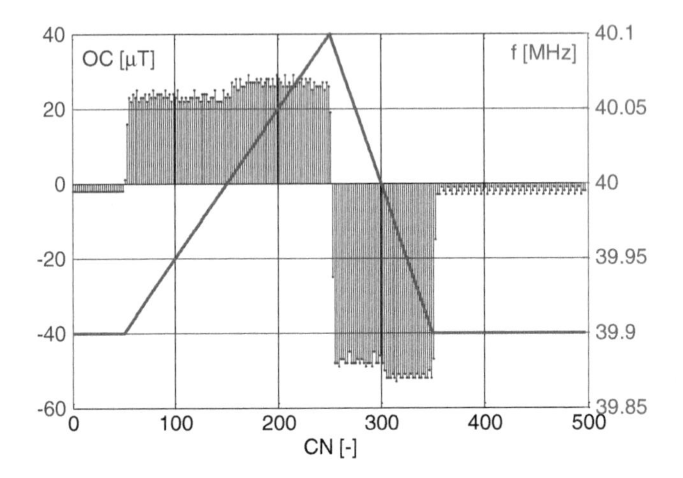

Fig. 6.17 Measured OC value for linear sweeping (node No. 4).

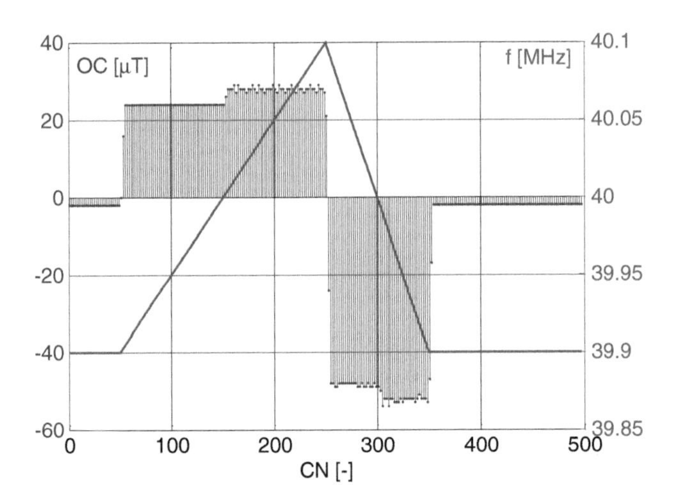

Fig. 6.18 Simulated OC value for linear sweeping (node No. 4).

(ECUs) connected together by the communication bus (CAN, FlexRay [3]...). ECUs usually come from different manufacturers. For robust and error free communication it is necessary to know internal parameters of each ECU. Nevertheless car manufacturers know only what is declared by ECU manufacturer and they need independent methods to validate the parameters. An algorithm for estimation of synchronization mechanism parameters can be verified using this model. The model is being used for validation of methods for

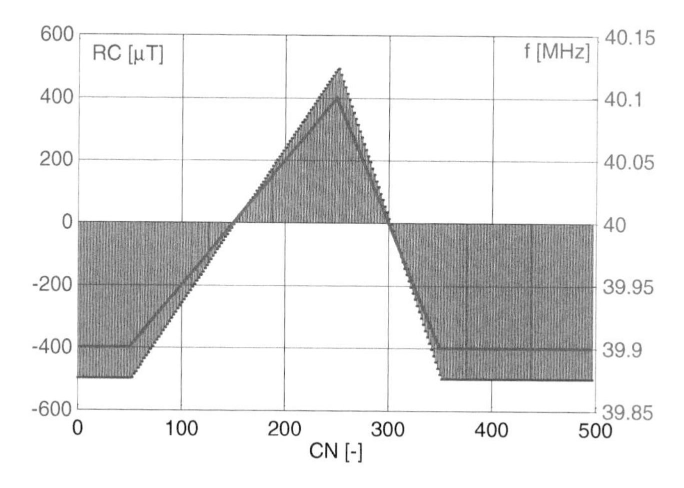

Fig. 6.19 Measured RC value for linear sweeping (node No. 4).

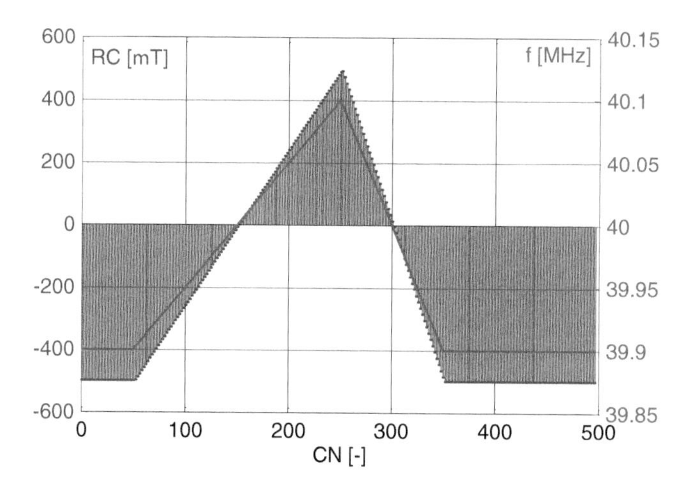

Fig. 6.20 Simulated RC value for linear sweeping (node No. 4).

measurement of FlexRay node parameters related to synchronization mechanism, e.g., *pClusterDriftDamping, pRateCorrectionOut.*

Consider following case. We would like to develop algorithm for evaluation the time of 1 μT of a tested FlexRay node. We try to send synchronization frames with small advance or delay and observe change in arrival time tested node frames placed in static slot. Correction can be caused by offset or rate correction. To minimization influence of rate correction (moreover influenced

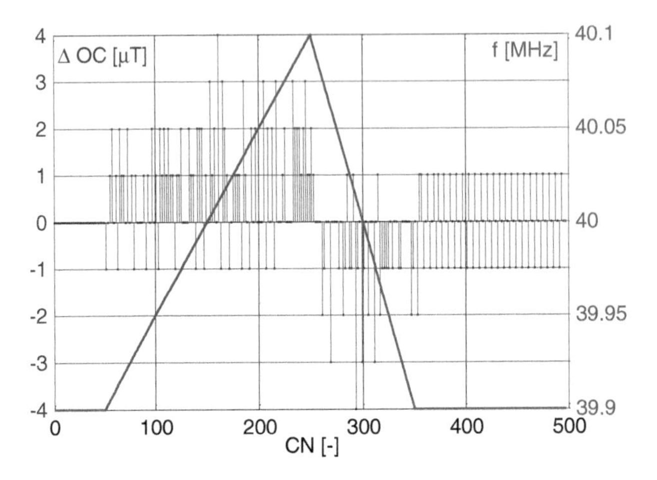

Fig. 6.21 Differences between measured and simulated OC value for linear sweeping (node No. 4).

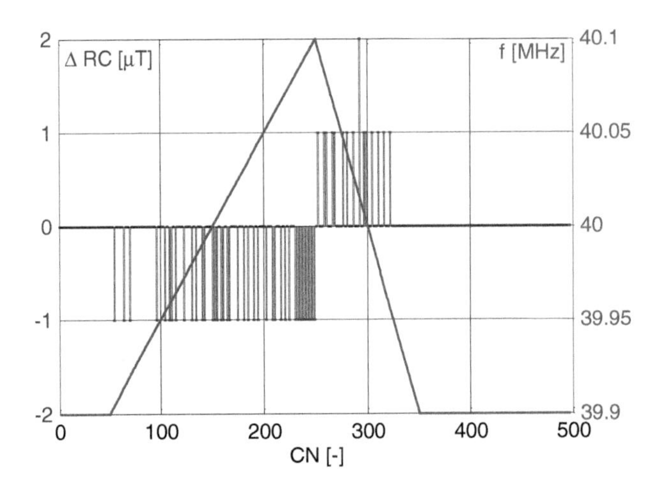

Fig. 6.22 Differences between measured and simulated RC value for linear sweeping (node No. 4).

by *pClusterDriftDamping*) it is possible to send shifted frames only in odd communication cycles. Now we would like to investigate our algorithm functionality. This is not possible with standard FlexRay hardware available on the market. FlexRay controller able to modify frame sending time and an instrument for precision time measurement (in ns range) are needed. But with model it is very easy to evaluate algorithm behaviour and not to waste time with complex experiments on real network.

Another application is a test in case of a fault [19] in synchronization mechanism. The investigation of behavior of FlexRay synchronization mechanism in case of a user-defined fault is practically impossible in the real system. Standard FlexRay controller is integrated in silicon chip. Thus, an easy and inexpensive possibility of user-defined error injection into the synchronization mechanism does not exist. However, it would be useful to know this behavior. This task is quite easy to do with the model and the behavior of the real system with a particular error occurring in the synchronization mechanism can easily be examined.

6.3.5 FPGA based FlexRay Node Tester

The issue mentioned in previous section is that standard FlexRay controllers available on market offer limited usability for testing purposes. For example it is not possible to change behaviour of start-up or synchronization mechanisms, control sending time for individual frames or manipulate with some internal parameters. One possible solution is to implement special FlexRay controller in VHDL language. The architecture of the FlexRay controller is shown in Figure 6.23. Motivation for this work is provide framework for implementation of an arbitrary method for FlexRay node parameter evaluation.

The FlexRay controller offers adjustable number of Rx buffers, filter of IDs, and an adjustable number of Tx buffers with extern or time stamp

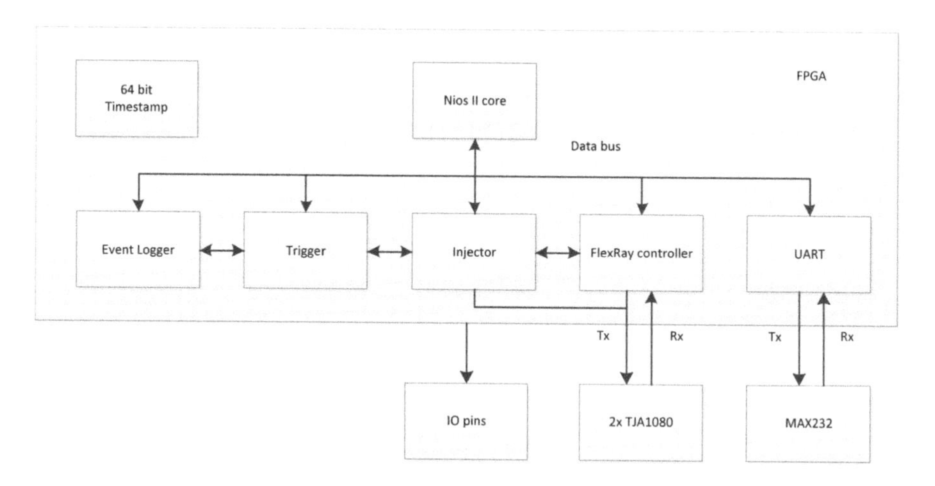

Fig. 6.23 FlexRay Bus Tester Architecture [17].

triggering. This controller can be used as a non-cold-start node, leading cold-start node, passive observer, or even as a unit disturbing the communication on the bus. The user have to decide which function to use and how to combine them together in order to test a specific parameter of node or cluster.

In future we would like to provide complex FlexRay node tester for industrial application. FlexRay Node Tester will be use for tests of FlexRay parametrization in automotive ECUs at car manufacturer site. The purpose of integration tests is evaluation that actual ECU parameters fit the specification.

6.4 Anti-lock Braking System ABS

Anti-lock braking system (ABS) is an independent system which ensures controllability of the vehicle in the braking moment when the wheels lose the adhesion to the road and try to block themselves. A blocked wheel may result in an uncontrollable skid, that means the vehicle is uncontrollable as well. The controllability is achieved through periodical decrease and increase of the brake fluid pressure at the individual wheels. The frequency of this intervention is between 10 and 16 Hz and the control algorithm passes several phases in every period.

Ensuring the controllability brings certain advantages as well as disadvantages. For example, a car without the ABS unit can stop much faster than a car with the ABS on a solid surface, but with considerable tyre damage. The braking distance can be shortened only on a slippery or loose surface.

The purpose of the braking algorithm is to achieve the highest braking efficiency while maintaining good controllability. It was proven, that this range is around 10% of the wheel slip. Thus the ABS algorithm is trying to keep the wheel at this slip value by regulating the brake fluid pressure during the braking cycle.

In a real conditions it often happens that some wheels have to brake on a loose surface, while the others brake on a good dry tarmac. Under these conditions — if also the conditions for maximal braking on the individual wheels are kept — the vehicle is oversteering. In order to decrease oversteering, various methods of interconnected regulation of the pressure between the wheels are used. Unfortunately, this effect causes an extension of the braking distance.

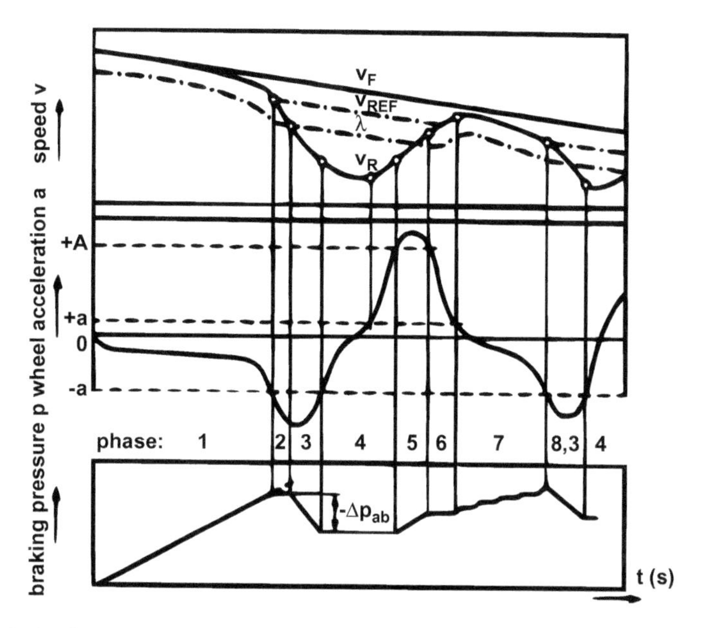

Fig. 6.24 Cycle of the main algorithm of the ABS unit: v_F — vehicle speed, v_{REF} — reference speed, v_R — wheel circumferential speed, λ — slip threshold, A, a — wheel circumferential acceleration (deceleration) threshold, Δp_{ab} — decrease of braking pressure [25].

6.4.1 Description of Main Algorithm of ABS

As already mentioned, the algorithm of the anti-lock braking system passes eight phases during one cycle. This cycle is depicted in Figure 6.24 and described more detailed in [7] and [5].

Phase 1: This phase occurs only once — before the intervention of the ABS unit. The vehicle behaves like there was no anti-lock braking system installed in it. So far a bigger slip does not occur, there is no need to adjust the braking pressure, so the braking cycle stops in this phase. The pressure in the brakes of the particular wheels corresponds with the pressure created by the braking pedal and strenghtened by the brake booster. But if the brake pressure grows till the wheel circumferential acceleration (deceleration) reaches the value $-a$, the ABS unit activates itself and switches to phase 2. At the end of this phase the current wheel speed, considered as the reference speed v_{REF} (the speed from which the current vehicle speed is derived), is saved.

Phase 2: This phase occurs only once as well — in the first braking cycle of the main algorithm of the ABS unit. In this phase, the constant pressure is kept until the wheel speed decreases below the speed defined from the slip threshold. This is achieved by closing the electromagnetical valve, which conveys the pressure into the given wheel. As soon as the wheel speed is lower, the main algorithm switches into phase 3.

Phase 3: In this phase the wheel speed decreased so much, that it is necessary to rotate it to the optimal slip range again. That means to the range, where braking is most effective and a good controllability of the vehicle is guaranteed. This is achieved when the electromagnetical valve is open. This valve then sends the brake fluid from the brake cylinder into the container incorporated in the ABS construction. This fluid is then progressively pumped back into the brake circuit, between the pedal and the ABS unit. This causes the known kicks in the brake pedal when the ABS unit is activated. The brake fluid is pumped back into the braking circuit, otherwise the brake pedal would be dropped so much that further braking would not be possible as all the liquid would be placed in the container. As soon as the wheel circumferential acceleration (deceleration) threshold reaches the value $-a$ again, the main algorithm of the ABS unit passes into phase 4.

Phase 4: Constant pressure is kept in this phase as well. The constant pressure is set, because the brake system has a certain delay and the decreased pressure in the given wheel is expressed in a certain time. If the pressure would be decreased (phase 3) until the wheel would rotate at the speed of the vehicle, we would reach the saturation point and the braking efficiency would be significantly worsened. As soon as the wheel circumferential acceleration (deceleration) threshold reaches the value A, the main algorithm passes into phase 5.

Phase 5: The pressure is increased in this phase again. Thus the discharge valve is closed and the inlet valve is open for a short moment until the wheel circumferential acceleration (deceleration) threshold value decreases to the value A again.

Phase 6: In this phase the wheel is in the most effective braking phase, so the vehicle is very controllable — the slip value is around 10%. The pressure is thus kept constant, until the wheel circumferential acceleration (deceleration) threshold value decreases below the value a.

Phase 7: Now the wheel is rotating almost at the speed of the vehicle. The vehicle is very controllable at this moment, but the braking effects are almost none, so the pressure in the brake cylinder of the given wheel is increased due to periodical opening of the inlet valve. As soon as the circumferential acceleration (deceleration) of the wheel is equal to $-a$, the current wheel speed is recorded as the reference speed v_{REF}, which is important for the calculation of the current vehicle speed. Now the phase 8 starts.

Phase 8: Phase 8 is basically a name for the phase 3 of the second cycle. The whole main algorithm of the control unit is thus repeated from phase 3 to phase 7 at the frequency of 10–16 times per second.

6.4.2 Description of Mechanical Construction of ABS

In the anti-lock braking system it concerns primarily two parts — 12 electromagnetical valves and one electromagnetic motor propelling the pumps of the brake fluid. The configuration of these parts is depicted in Figure 6.25.

The control unit of the anti-lock braking system ABS 8.0 has — in comparison with the older versions — four more electromagnetical valves (No. 3 and 4). That means instead of eight valves (No. 9 and 10 see Figure 6.25), which are enough for the principle of the ABS unit, there are twelve valves now.

In a very simplified way one could say that in normal condition the valves No. 9 are open and the valves No. 10 closed. If one of the wheels is in slip, the respective valve No. 9 of the given wheel closes and prevents further pressure rise. The pressure in the given wheel is kept at a constant level for a short moment, and if the wheel does not start rotating again, the valve No. 10 is open and thus the pressure in the respective brake cylinder decreases, so the wheel can start rotating again. For a more detailed description of controlling these valves, see Section 6.4.1.

If the pressure in the brake cylinder of the given wheel is decreased through the valve No. 10, the brake fluid from the given wheel accumulates in the pressure container 8, from which it is pumped by a pump 7 moved by an electromagnetic motor 6 into the braking system between the pedal and the ABS unit.

The other mentioned electromagnetical valves 3 and 4 are used for additional functions ensured by the ABS control unit, e.g., ESP and ASR

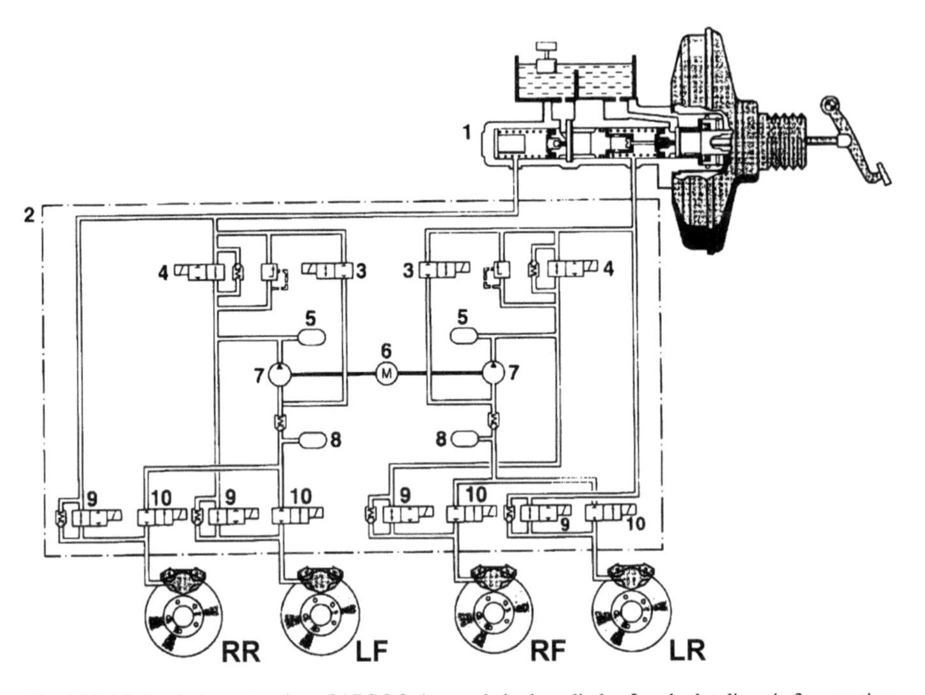

Fig. 6.25 Mechanical construction of ABS 8.0: 1 — main brake cylinder, 2 — hydraulic unit, 3 — suction valves, 4 — bypass valve, 5 — damping chamber, 6 — electromagnetic motor, 7 — drain pump, 8 — pressure accumulator, 9 — inlet valves, 10 — discharge valves, RR — right rear wheel, LF — left front wheel, RF — right front wheel, LR — left rear wheel [25].

systems etc. However, the analysis and diagnostics of these systems is not a topic of this chapter.

6.4.3 ABS development platform

System Architecture

The system architecture is depicted in Figure 6.26. There are four ECUs acting as wheel rotation sensors. In a simulation mode the ECU measures frequency generated by a two channel function generator. It allows to generate signal with additional noise to simulate a more realistic behavior. In operating mode a real sensor of wheel rotation is connected in current loop and thus a short-circuit can be detected. The actual wheel rotation speed is then transmitted over the FlexRay network in a static segment. For our purpose we use the following setting based on the BMW network design [4]. The communication

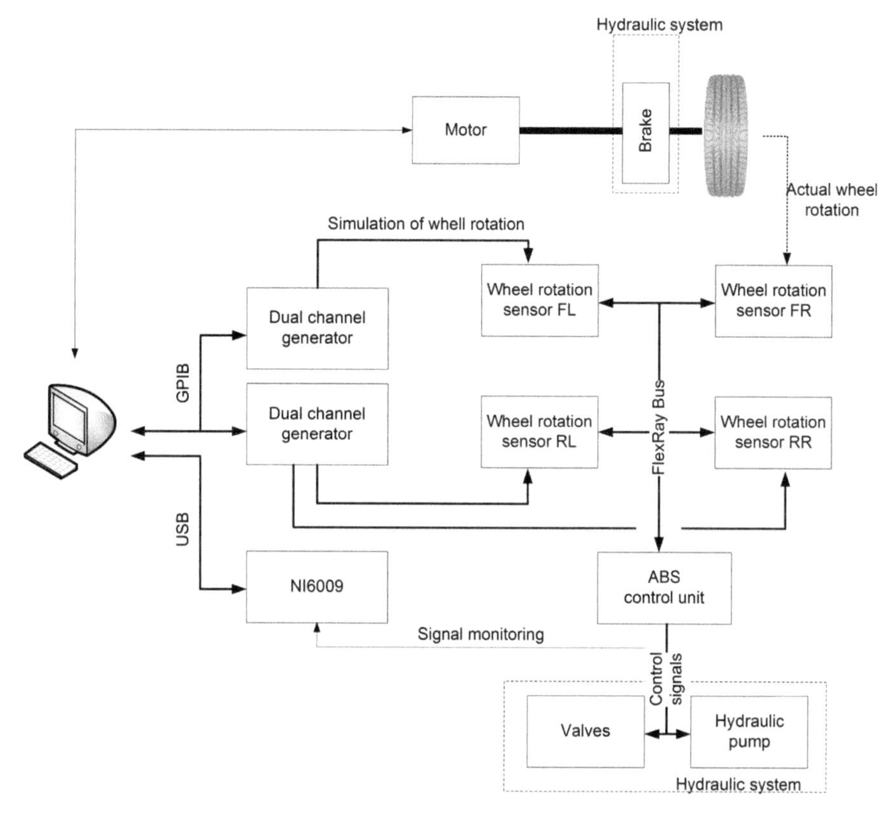

Fig. 6.26 ABS system architecture.

cycle duration is equal to 5 ms where merely static segment is used. There are 75 static slots with static slot duration of 0,07 ms. The frame payload is 128 bits. Currently, each ECU uses 4 static slots for data transmission and thus the maximum transmission delay is 1 ms due to FlexRay cycle length. To minimize bandwidth data are transmitted just via one channel. However, the transmission delay can be shortened by using more static slots by one ECU. The software in the central node realizes anti-lock braking control function described in [8]. Moreover, the central ECU receives acceleration in two axis from a sensor located on the rotated wheel. Based on acceleration, the wheel rotation speed is calculated and thus it can serve as a backup line if a short-circuit is detected in the current loop of the main sensor. The acceleration and wheel speed is transmitted to the central node via bluetooth network due to simple mechanical realization.

Hardware Description

The ECUs were designed as a modular system. The main board of each single node consists of a 16-bit microcontroller MC9S12XF by Freescale with a XGATE co-processor and integrated FlexRay controller, two bus drivers TJA1080 by Philips and a Field-programmable Gate Array Spartan-3E by Xillinx for a FlexRay bus guardian block. The central ECU contains also an expansion board with control logic for valves and high-power circuits for hydraulic pump. The design of ECUs include low-power mode and wake-up via FlexRay network. The wheel ECUs with integrated tyre pressure, temperature and acceleration sensors are based on the MPXY8300 microprocessor by Freescale. The radio frequency (RF) output is also integrated on a chip. The wireless sensor is powered by a long-life battery located on the board.

6.4.4 Vehicle model

Half-car model used in [24] was used for proposed ABS system validation. We assume that the car is driven on a rugged surface which is simulated by white-noise. Due to the fact that wheel speed is not directly monitored by ABS unit, but it is transmitted over FlexRay network, the transmission of speed information to the ABS central unit is discontinuous. As mentioned above, each wheel ECU uses four static slots for data transmission. In order to model this situation, actual wheel speed is sampled with period 1 ms. Moreover, transport delay 70 μs is added.

In order to model car speed data transmission from accelerometer, sampling and transport delay is also added to this signal. To keep a more realictic situation, white-noise is also added to actual wheel speed.

6.4.5 Proposed ABS system

The 8-step ABS algorithm [5, 8] was used as a reference ABS model. This control algorithm measures an actual car speed at the beginning of the cycle. In the rest of the control cycle it only predicts the actual car speed. This leads to rather high difference from the optimum slippage of about 10%. Newly designed algorithm uses information from additional accelerometers for precise slippage evaluation during the whole control cycle. If the slippage stays below the 20% limit, our algorithm works like the standard one. When this limit is exceeded, control algorithm goes into so called emergency mode,

Table 6.2. Brake distance.

	Conventional ABS system	Our proposed ABS system
Dry asphalt	41,14 m	42,85 m
Packed snow	99,73 m	100,1 m
Ice	290,5 m	275,1 m

where the braking pressure for the affected wheel is quickly reduced to return the slippage close to the optimal 10% value.

Proposed ABS system was validated on vehicle model[24] on dry asphalt, packed snow and ice. Simulation results are shown in the Figure 6.29. As it can be seen, a car is decelerating from 100 km/h. Table 6.2 shows braking distance of conventional 8-step ABS and our proposed ABS system. For dry asphalt and packed snow our system is a little bit worse in braking distance, however the car is better controllable.

6.4.6 System Validation

The whole concept is verified in the following phases. Firstly, the potential of shortening the braking distance was verified on a simple vehicle model. Secondly, the concept was implemented to a 3D vehicle model (see Figure 6.27.) and the results were evaluated for road surfaces of different

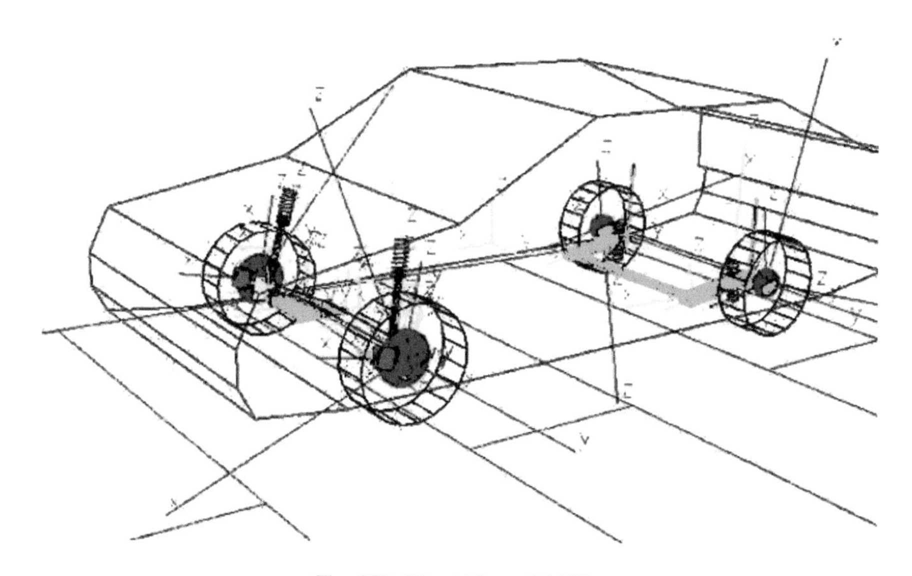

Fig. 6.27 3D vehicle model [20].

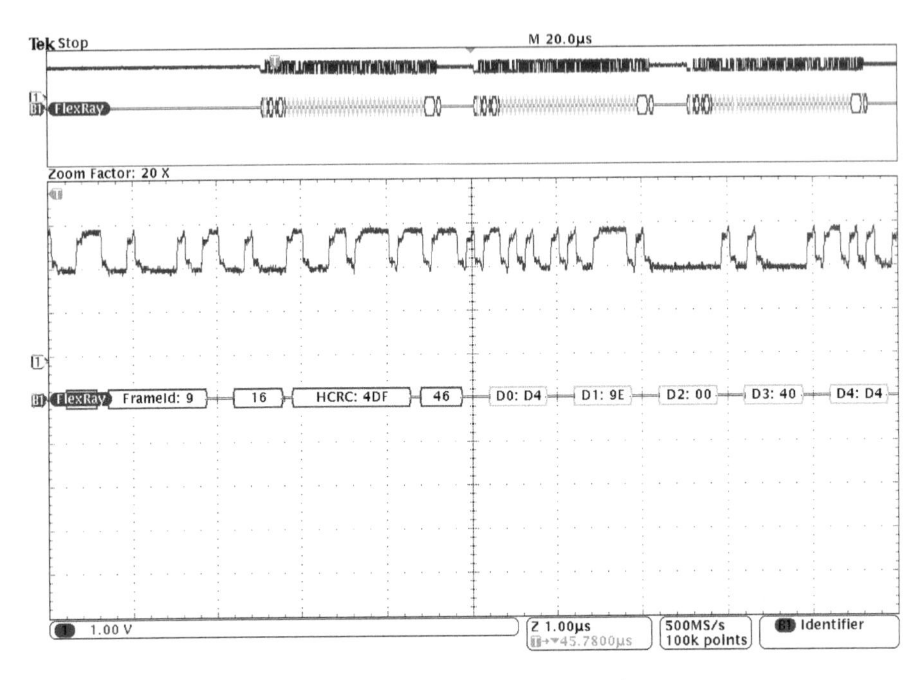

Fig. 6.28 FlexRay communication snapshot.

profiles, such as smooth profile, deterministic sinusoidal profile or stochastic excitation profile. As the last step, the algorithm is implemented in a developed ABS platform. Figure 6.28 shows part of the communication on one FlexRay channel.

The vehicle model is equipped with several suspension types. The reference model has passive suspension and the control models have either semi-active dampers or limited active actuators. The brake circuit is extended by an anti-lock controller based on an ABS algorithm from [8]. The 3D vehicle is designed in a multibody package SIMPACK; however, the controller of semi-active dampers as well as the anti lock algorithm is implemented in a control engineering tool MATLAB/Simulink. The both software packages are connected by means of co-simulation based on interprocess communication [23]. Since the anti-lock controller induces significant frequencies to the system, one must pay attention to selection of proper tyre models [12].

Final results in Table 6.2 show, that at the 10% slippage providing optimal directional controllability of the vehicle, the braking distance increases

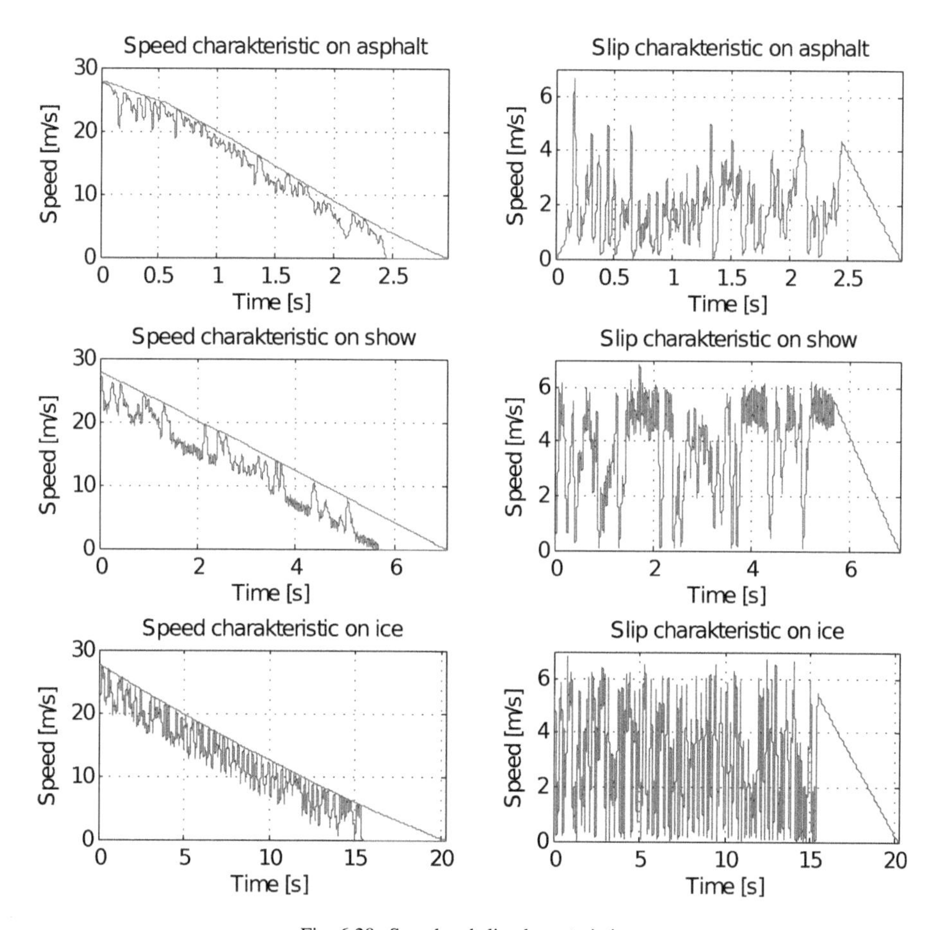

Fig. 6.29 Speed and slip characteristics.

on the dry and solid surface. Only for the slippery road surface (like ice) the braking distance is shorter than for the standard control algorithm. This effect of ABS system is well known and for the new algorithm it applies even more. Nevertheless, the described simulations show that the correct application of the FlexRay communication channel has no or negligible influence on ABS performance and it can be used as a replacement for current loops with pulse width modulation that are used today. Using the FlexRay allows to share the same communication infrastructure (and available information) by other control systems in vehicle and thus it offers possible improvements of particular control algorithms.

6.5 Conclusions

In the chapter the emerging FlexRay standard for in-vehicle communication is introduced quite in detail. In the first part of the chapter the mechanisms important for FlexRay correct functionality are described, especially the synchronization mechanism. Its analytic model is then derived and its validation is presented. Plausibility of the model was validated by comparison of behaviour of the real FlexRay system consisting of 10 network nodes and the simulated system with the same parameters under the same conditions. The model was validated using two scenarios. The static scenario uses the constant deviation of a node oscillator frequency from the nominal value while the dynamic scenario is based on continuous changes of the oscillator frequency in the selected node. In both cases the real system and the model behaves in the same way and model can be declared as plausible. Finally the simple example of the model application is presented and future intentions of its use are mentioned.

The second part of the chapter demonstrates the implementation of new FlexRay communication protocol into the anti-lock braking system. It introduces the developed ABS platform consisting of five FlexRay nodes that will be used for enhanced ABS control algorithms verification. The system is designed for simple integration together with other x-by-wire systems. Acceleration sensor was added and — based on the measured wheel acceleration — the conventional ABS algorithm was modified to keep slip rate below the 20% level. Improved ABS model is simulated on a simple vehicle model and then on a 3D vehicle model. Simulation shows that proposed algorithm achieves better results on a surface covered by ice. We assume that the proposed system will be tested also on the quarter car experimental stand developed at the Laboratory of Mechatronics and Dynamics of the Josef Bozek Research Center on Combustion Engines and Automotive Engineering [20].

References

[1] Amos Albert. Comparison of event-triggered and time-triggered concepts with regard to distributed control systems. In *Embedded World 2004*, pp. 235–252, 17.19.02.2004 2004.

[2] Eric Armengaud and Andreas Steininger. Remote Measurement of Local Oscillator Drifts in FlexRay Networks. In *DATE: 2009 DESIGN, AUTOMATION & TEST IN EUROPE CONFERENCE & EXHIBITION, VOLS 1-3*, Design, Automation and Test in Europe Conference and Expo, pp. 1082–1087, 345 E 47TH ST, NEW YORK, NY 10017 USA,

2009. IEEE. Design, Automation and Test in Europe Conference and Exhibition, Nice, FRANCE, APR 20-24, 2009.

[3] Eric Armengaud, Andreas Steininger, and Martin Horauer. Towards a systematic test for embedded automotive communication systems. *IEEE Transactions on Industrial Informatics*, vol. 4, no. 3, pp. 146–155, Aug. 2008.

[4] Josef Berwanger, Martin Peteratzinger, and Anton Schedl. FlexRay startet durch — FlexRay-Bordnetz fr Fahrdynamik und Fahrerassistenzsysteme, 2008.

[5] Bosch. *Automotive Handbook*. Society of Automotive Engineers, 4th edition, 1996.

[6] FlexRay Consortium. FlexRay protocol specification v2.1 rev. a, 2005.

[7] T.D. Day and S.G. Roberts. A simulation model for vehicle braking systems fitted with abs. 2002.

[8] Terry D. Day and Sydney G. Roberts. A simulation model for vehicle braking systems fitted with ABS. *SAE Technical Paper 2002-01-0559*, 2002.

[9] Armengaud E., Rothensteiner F., Steininger A., Pallierer R., Horauer M., and Zauner M. A structured approach for the systematic test of embedded automotive communication systems. In *Test Conference, 2005. Proceedings. ITC 2005. IEEE International*, 2005.

[10] Klaus Echtle and Soubhi Mohamed. Clock Synchronization Issues in Multi-Cluster Time-Triggered Networks. In MullerClostermann, B and Echtle, K and Rathgeb, EP, editor, *Measurement, Modelling, and Evaluation of Computing Systems and Dependability and Fault Tolerance*, volume 5987 of *Lecture Notes in Computer Science*, pp. 39–61. Univ Duisburg Essen, Inst Comp Sci & Business Informat Syst, 2010. 15th International Conference on Measuring, Modeling and Evaluation of Computing Systems/Dependability and Fault Tolerance (MMB-DFT), Essen, Germany, MAR 15-17, 2010.

[11] K. Jang, I. Park, J. Han, K. Lee, and M. Sunwoo. +Design framework for FlexRay network parameter optimization. *International Journal of Automotive Technology*, vol. 12, no. 4, pp. 589–597, Aug. 2011.

[12] S. T. H. Jansen, P. W. A. Zegelaar, and H. B. Pacejka. The influence of in-plane tyre dynamics on ABS braking of a quarter vehicle model. *Vehicle System Dynamics*, vol. 32, pp. 249–261(13), Numbers 2–3/August 1999.

[13] J. Malinský and J. Novák. Verification of FlexRay Start-up Mechanism by Timed Automata. *Metrology and Measurement Systems*, vol. 17, no. 3, pp. 461–480, 2010.

[14] Jan Malinský. *Intrusive Tests in FlexRay Standard*. PhD thesis, Czech Technical University in Prague, 2010.

[15] N Navet, YQ Song, F Simonot-Lion, and C Wilwert. Trends in automotive communication systems. *Proceedings of the IEEE*, vol. 93, no. 6, pp. 1204–1223, Jun 2005.

[16] J. Novák. Testing of CAN Based Automotive Distributed Systems Using a Flexible Set of IP Functions. *Sensors and Transducers*, vol. 8, no. Special, pp. 54–64, 2 2010.

[17] Martin Paták. Methods for testing the flexray start-up mechanism. Master's thesis, CTU in Prague, 2012.

[18] Yasser Sedaghat and Seyed Ghassem Miremadi. Investigation and Reduction of Fault Sensitivity in the FlexRay Communication Controller Registers. In Harrison, MD and Sujan, MA, editor, *Computer Safety, Reliability, and Security, Proceedings*, volume 5219 of *Lecture Notes in Computer Science*, pp. 153–166, HEIDELBERGER PLATZ 3, D-14197 Berlin, Germany, 2008. EWICS TC 7; Ctr Software Reliabil; Newcastle Univ; Warwick Med Sch; AdaCore; ReSIST; Qinetiq; Adelard; TTE-Syst; British Comp Soc; IFIP; DECOS; Austrian Comp Soc; Gesell Informatik e V; ENCRESS, SPRINGER-VERLAG

BERLIN. 27th International Conference on Computer Safety, Reliability, and Security, Newcastle upon Tyne, ENGLAND, SEP 22-25, 2008.

[19] Yasser Sedaghat and Seyed Ghassem Miremadi. Classification of Activated Faults in the FlexRay-Based Networks. *Journal of Electronic Testing-Theory and Applications*, vol. 26, no. 5, pp. 535–547, Oct 2010.

[20] Pavel Steinbauer, Zbynek Sika, Michael Valasek, and Pavel Mikolas Tomas Skopalik. HIL experiments with quarter car. *Inzenyrska mechanika*, pp. 99–106.

[21] Gang-Neng Sung, Chun-Ying Juan, and Chua-Chin Wang. Bus guardian design for automobile networking ecu nodes compliant with flexray standards. In *Consumer Electronics, 2008. ISCE 2008. IEEE International Symposium on*, pp. 1–4, April 2008.

[22] Bogdan Tanasa, Unmesh D. Bordoloi, Petru Eles, and Zebo Peng. Scheduling for fault-tolerant communication on the static segment of flexray. In *Proceedings of the 2010 31st IEEE Real-Time Systems Symposium*, RTSS '10, pages 385–394, Washington, DC, USA, 2010. IEEE Computer Society.

[23] Ondrej Vaculin, Wolf-Reiner Krueger, and Martin Spieck. Coupling of multibody and control simulation tools for the design of mechatronic systems. In *ASME Design Engineering Technical Conference 6 A, Proceedings of*, pages 199–206. ASME, 2001.

[24] Michael Valasek, Ondrej Vaculin, and Jaromir Kejval. Global chassis control: Integration synergy of brake and suspension control for active safety. 2004.

[25] F. Vlk. *Automobilova elektronika: Asistencni a informacni systemy*. Number sv. 1. Frantisek Vlk, 2006.

7

Implementation of a CO Concentration Monitoring System using Virtual Instrumentation

Raul Ciprian Ionel*, Aurel Gontean, and Patricia Gherban-Draut

*University "Politehnica" from Timişoara, Bd. Vasile Parvan, Nr. 2, 300223, Timişoara, Romania; *raul.ionel@etc.upt.ro*

Abstract

Different types of gas sensors are used to implement applications which investigate ambient air pollution levels. Carbon monoxide (CO) concentration measurement is an integral part of dedicated environment monitoring stand-alone specialized systems. This chapter presents an alternative solution for CO concentration monitoring, based on an original virtual instrumentation concept. The advantages of the proposed application include data logging,statistical calculations, remote access or software and hardware flexibility. Comparative experimental results are also provided.

Keywords: Virtual Instrumentation; Gas Sensor; LabVIEW; CO Concentration.

7.1 Introduction

The use of Virtual Instrumentation (VI) is becoming more and more popular when dealing with sensor based data acquisition and analysis. As a main goal,

Advanced Distributed Measuring Systems — Exhibits of Application, 163–182.

VI delivers the means to rapidly develop a competitive tool which meets the technical performances of traditional devices. Using customizable software and modular hardware, one can implement applications which are extremely powerful and complex, at the same time having lower costs. The technological innovations rhythm together with customer needs for increased functionality in smaller dimensions products, have caused the spread of VI in domains like: health and medicine, environmental monitoring, structural monitoring, clean energy production, data transmission, industrial control, education, robotics and/or automation.

Basically, a virtual instrument can be defined as a software and hardware ensemble which has the purpose of replacing a stand-alone, dedicated device. The main advantage is represented by the possibility to exploit the calculus power and performances of the PC on which the software component runs. Because of the ease with which one can create and deploy flexible user-defined test systems, VI extends to computerized applications for controlling certain processes using collected and analysed data. So the combination between PCs, Graphical Programming Languages and modular hardware has caused the emergence of a huge variety of virtual instruments.

Such solutions may demand the use of additional elements like sensors and built in signal conditioning. Accelerometers, strain gauges, load cells, thermocouples and thermistors are just a few types of sensors which are used in measurement and automation application. Voltages and currents can be acquired or generated using available hardware. For a data acquisition system, modularity becomes extremely important since application requirements may change as time passes. Dedicated modules allow direct sensor connectivity and hot-swapping. These options become an advantage if the user decides to change modules (thus, application functions) while the system is powered on.

The evolution of VI allows users to take advantage or wireless data transmission and remote monitoring using the Internet or other communication network (GSM for example). Data acquisition using connections like Ethernet, USB, PCI and PCI Express, are provided in order to meet the newest user challenges. It all comes down to user preferences and application demands. For example, if data transmission speed is the critical aspect when implementing an industrial application, one can choose to use data acquisition hardware which communicates over the Ethernet. The advantage is represented by the transmission speed of up to 1 Gb/s and distributed I/O measurement possibilities [1–5, 16].

Finally, a virtual instrument incorporates three essential features.

- Intuitive software tools used for applications development.
- Modular I/O based on state of the art commercial technologies
- A PC platform for synchronization, accuracy and increased performances.

When comparing VI measurement solutions against the performances offered by stand-alone devices, there are a few advantages which one can take into account:

- Flexibility and adaptability (VI can be rapidly adapted to application changes).
- The existence of a large palette of hardware modules which facilitate applications development
- The support offered by VI producers, the enormous amount of written electronic documentation, publications, videos or other materials, provide the means to implement extremely efficient systems.

7.2 The Application — Carbon Monoxide Concentration Monitoring

First Starting from the previously presented concepts, this chapter presents an alternative solution for Carbon Monoxide (CO) concentration monitoring, based on an original virtual instrumentation concept. The advantages of the proposed application include data logging, statistical calculations, remote access or software and hardware flexibility. Comparative experimental results are also provided.

Carbon monoxide is a colourless, odourless, tasteless and highly toxic substance. Measurements of CO concentration can be found in diverse domains, including the study of gases which are emitted by car engines. CO is not only a pollutant component, but its presence is a proof of uncompleted combustion, generating efficiency losses and thus higher fuel consumption. Its concentration is similar to that of most gases and can be measured with adequate sensors. Technical validation for engine functionality is not approved, if the CO concentration in flue gases exceeds established limits (by law). CO concentration monitoring in enclosed spaces becomes extremely important since a high enough level (exceeding 30-50 ppm) may cause headaches, fatigue, sensations of weakness or nausea. The use of dedicated sensors, including alarm circuits, is the recommended approach for preventing such situations [6–8].

Figure 7.1 presents the general structure of a virtual instrument used for data acquisition and processing. The proposed CO concentration monitoring solution is based on this VI architecture.

Sensors will convert a physical phenomena into signals which can be measured with appropriate hardware. Depending on the particularities of the application, signal conditioning is required in order to remove unwanted noise, amplify signals or excite the sensors. Data acquisition hardware is used to convert analog signals to digital signals which are processed by the computer.

An overview of the CO measurement system is illustrated in Figure 7.2.

The solid state TGS 2442 gas sensor is the core of the application. The sensor's functioning principle is based on the sensibility of metal oxides to different gases. Variations of the electrical parameters for the TGS 2442 sensing

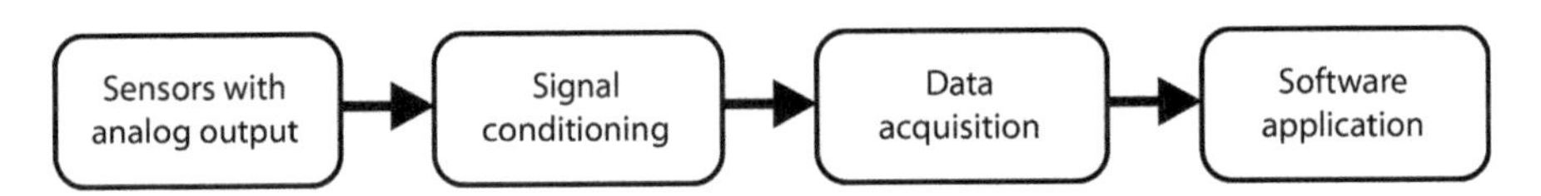

Fig. 7.1 Components of a virtual instrumentation used for data acquisition and analysis.

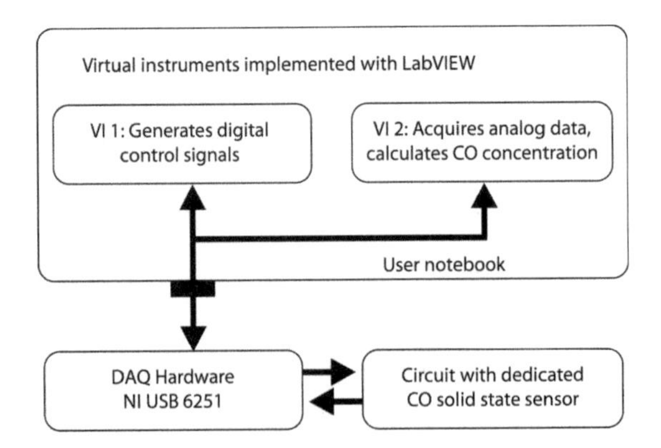

Fig. 7.2 Overview of the proposed CO concentration measurement system.

element are very consistent with concentrations of CO in the surrounding environment. By using data measured in controlled environments, it is possible to create a mathematical relation between the sensor's variable internal resistance and the volumetric concentration of carbon monoxide. This sensor represents a low cost, reduced physical dimensions and lower precision solution and is ideal for gas detecting. For this type of sensors a signal conditioning circuit is necessary and fulfils command tasks, temperature and humidity compensation [9–11, 17].

The USB-6251 board, produced by National Instruments, serves as data acquisition hardware. This device was considered as appropriate since the application requires both hardware generation of digital signals and analog data sampling. Digital signals are used for control purposes while analog data received from the measurement circuit is sampled and analyzed. A connector with screw terminals (CB-68LPR) was used to easily access individual signals.

The application for CO concentration measurement was divided into three main parts: acquisition, analysis, and presentation of data. It was programmed in National Instruments' LabVIEW development environment and was tested using a notebook computer with Dual Core — 1.8 GHz CPU, 2 GB DDR3 and Internet Connection. Interaction between the software component and the USB-6251 is performed by the NI-DAQmx driver.

7.3 Carbon Monoxide Measurement Method

As stated, the solid state TGS 2442 gas sensor, produced by Figaro Engineering, was used for measuring the CO concentration. This sensor provides low power consumption, increased CO sensitivity, reduced humidity dependence and reduced size.

A glass thermal insulation is placed between a ruthenium oxide (RuO_2) heater and an alumina substrate. A pair of heater electrodes is placed on the thermal insulation. The gas sensing layer, which is formed of tin dioxide (SnO_2), is printed on an electrical insulation layer which covers the heater. A pair of electrodes, for measuring the sensor resistance, is placed on the electrical insulator. Activated charcoal is filled between the internal cover and the outer cover for the purpose of reducing the influence of noise gases. The functioning principle is specific to semiconductor sensors. The target gas passes through filters placed on the upper side of the capsule and comes in contact

Fig. 7.3 The TGS 2442 carbon monoxide sensor.

with the tin dioxide. The CO molecules react with already absorbed oxygen molecules. This increases the number of free electrons in the sensing material and alters the conductivity of the material. As a consequence, the material resistance changes proportionally to the target gas concentration variations.

Figure 7.3 presents the actual TGS 2442 sensor capsule [11].

According to producer specifications, the typical detection range of the sensor is 30 ppm to 1000 ppm. The characteristic response for standard test conditions is presented in Figure 7.4 and was obtained from the producer technical notes. One can notice the evolution of the sensor internal resistance ratio with CO concentration. The R_S/R_0 quantity represents the sensor response for various gas concentrations. R_S is the internal resistance which changes as measurement time passes while R_0 is the sensor specific resistance at 100 ppm, obtained under standard experimental conditions. As CO concentration increases, the value of the internal resistance decreases [11].

The basic sensor structure, measuring circuit and operation signals diagram are graphically represented in Figure 7.5. For this particular application, the data acquisition device must provide 2 analog inputs and 3 digital output lines, plus a digital trigger for analog acquisition. The multi-function NI USB-6251 device offers the required features.

The sensor is heated through digital pulses programmed in the application and generated by the data acquisition device. Its output is measured and processed by a second virtual instrument that determines the concentration of carbon monoxide. This principle is consistent with the architecture presented in Figure 7.2.

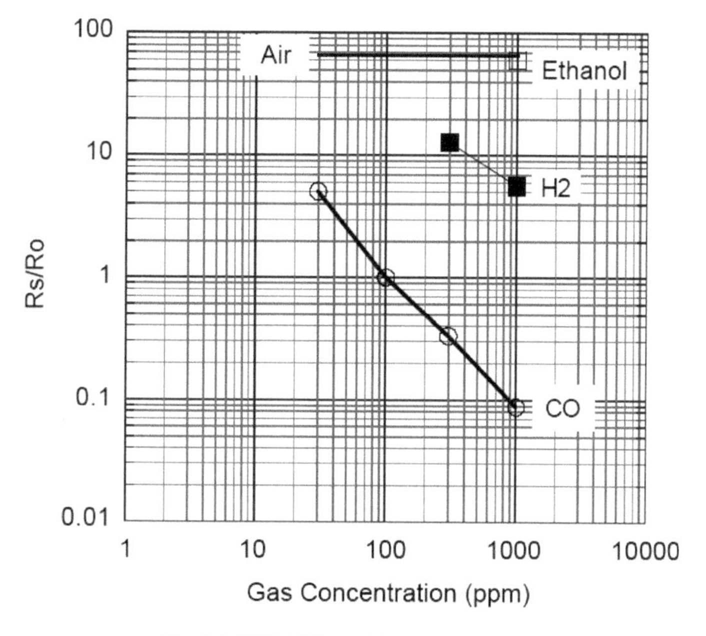

Fig. 7.4 TGS 2442 sensitivity characteristic.

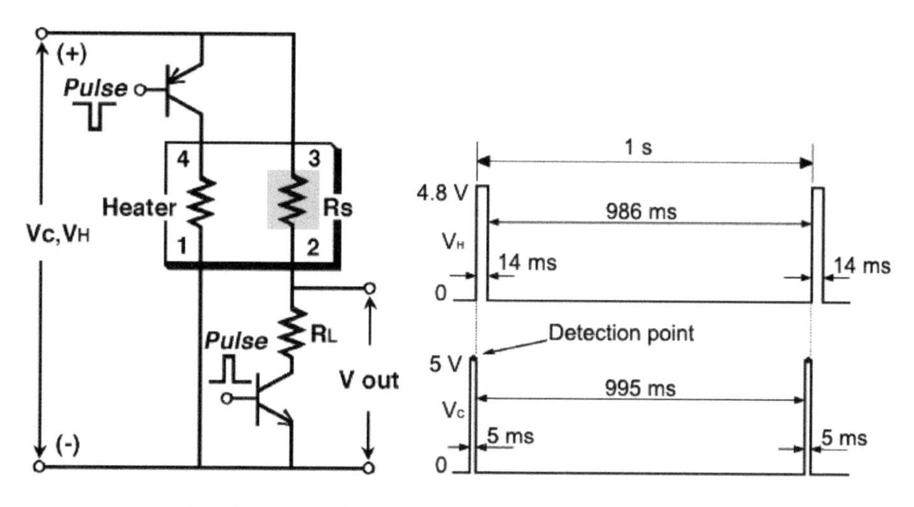

Fig. 7.5 Basic TGS 2442 structure, circuitry and operation signals.

One measurement cycle lasts 1 s and is composed of a heating cycle and a voltage acquisition cycle. The sensor's heater (connected between pins 1 and 4) requires a 4.8 V pulse applied for 14 ms, followed by a 0 V signal for the remaining 986 ms. A 200 mA current is required through the heater and was supplied by a BD140E power transistor. The sensing element (connected between pins 2 and 3) requires a 0 V signal for 995 ms followed by a 5 ms, 5 V pulse. The detection point where the data sampling must take place is timed at exactly 997.5 ms from the start of the measurement cycle.

Two more circuits for malfunction detection and temperature/humidity compensation were implemented. The timing of required signals implies a great deal of attention. Synchronization at millisecond level is necessary in order to provide correct functionality. This may cause several problems since the operation of both data acquisition device (which generates and records data) and computer programs needs to be tuned to perfection.

Several operation tests (for both the hardware and software components) were conducted with the sole purpose of studying if it is possible to obtain the timing relations indicated by the producer specifications. Figure 7.6 presents such an experiment, when the relation between the signals Pulse and RL_1 was tested using a digital oscilloscope. The Pulse signal was in the heating cycle while the RL_1 signal was used for concentration calculation. On the oscilloscope screen the logic of the signals is reversed due to particularities of the hardware implementation but the timing is in accordance with the application design specifications.

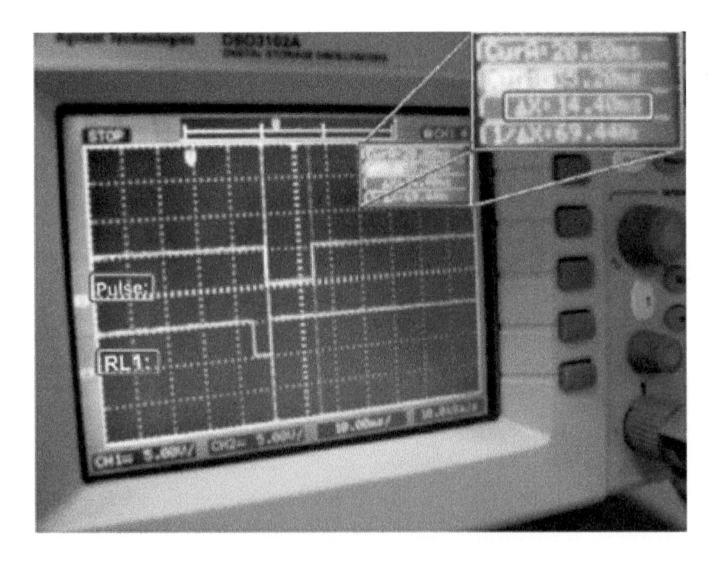

Fig. 7.6 Signal's synchronization testing: Pulse and RL_1.

The sensor's response stabilizes within minutes, depending on how long it has been inactive. Only when the acquired data is stationary can the correct concentration be calculated [12, 13].

The sensor's output determines the concentration of carbon monoxide according to Equation (7.1). This equation was implemented in the second virtual instrument (VI 2 according to Figure 7.2).

$$C = \left(\frac{V_{1\text{ref}} \cdot (5 - V_1)}{V_1 \cdot (5 - V_{1\text{ref}}) \cdot k_{\text{temp}}} \right)^{1/\alpha} \tag{7.1}$$

V_1 is the sensor output signal sampled at the moment of the Detection Point, as indicated in Figure 7.5. $V_{1\text{ref}}$ is the value of V_1 at a known CO concentration (typically 100 ppm) and a constant value of 1.5 was used. The k_{temp} is the coefficient for temperature compensation and can be obtained from a predefined table. For example, at 25°C (temperature in the room where the measurements were conducted), the coefficient has a value of 1. Testing of the correct formula calculation was initially performed using manual introduction of the room temperature. The inserted value was used by the application to automatically search the proper coefficient in a predefined table. The measure of the slope in the sensor's characteristic is defined by α [11, 14, 18]. For this particular calculation the value for α was set to -1.05 because of the specifications of the sensor capsule.

Figure 7.7 presents the implementation of the concentration of carbon monoxide calculation in VI 2. This is a small part of the code. The V_1 voltage samples are received from the NI USB-6251 while the k_{temp} is looked up in the predefined table. The Calculation Result is sent for further processing.

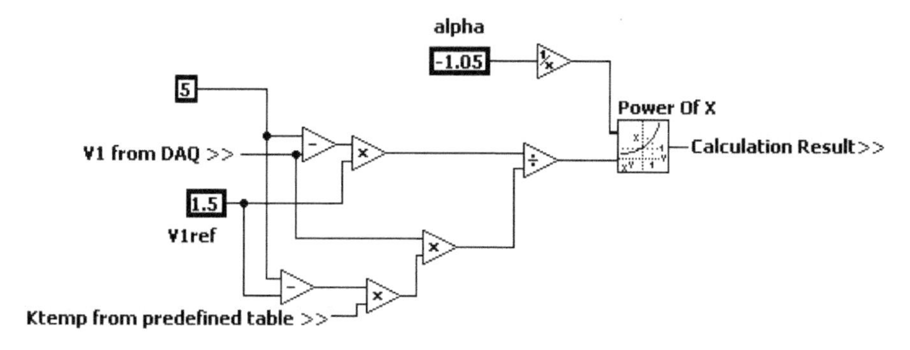

Fig. 7.7 Calculation of CO concentration in LabVIEW.

7.4 Virtual Instrumentation Hardware Functionality

The hardware generates digital signals programmed through software and receives analog data obtained from the CO measurement circuit. The purpose of the virtual instrumentation hardware is to exchange information between the computer and the sensor circuit. More specifically, the hardware generates the digital signals programmed through software and receives analog data obtained from the CO measurement circuit. A total of three digital lines from Port 0 (P0.0 through P0.2) and two analog input lines (AI0 and AI1) are reserved by the application. These lines together with the ground signal and the power supply are provided by the data acquisition device.

Figure 7.8 includes the basic schematics and the CO measurement circuit. In order to measure CO concentration, the communication between the software and hardware components needs to use both digital output signals and analog input lines. For this reason, the hardware has to be a multi-functional data acquisition device able to perform those tasks. The NI USB-6251 can acquire analog data on maximum 16 input channels, with up to 1.25 Ms/s and can generate both digital and analog signals. This can be done using three digital ports and two 16-bit analog output lines. Frequency and start time for analog acquisition can be supplied either by an internal clock (through software) or by an external trigger signal. For example, if the trigger is a digital signal, the acquisition can be initiated by its falling slope, as is the case in this application.

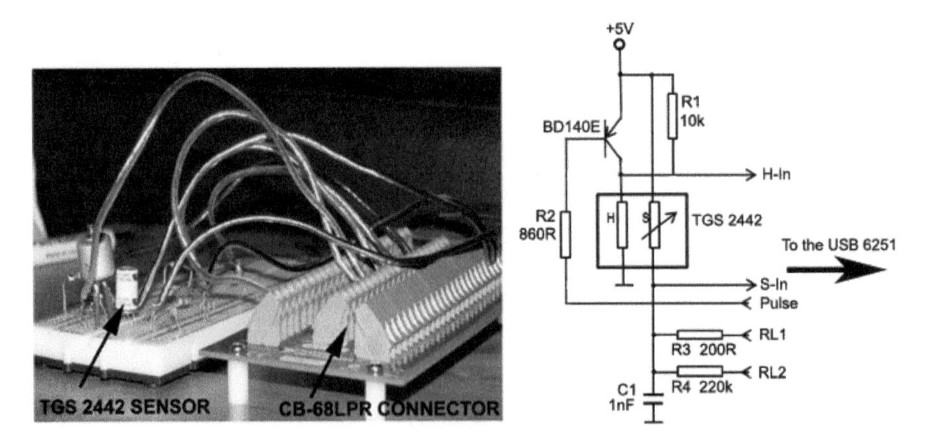

Fig. 7.8 Concentration measurement circuit and the corresponding basic schematics.

7.5 Virtual Instrumentation Software Functionality

The first virtual instrument (VI 1) generates the digital signals, out of which the falling slope of Pulse will be used to trigger the sampling of V_1 at the Detection Point. The signals which are fed to the digital output lines of the hardware device have the form of constant Boolean Arrays. A Wait function determines the number of ms for each state. These operations are performed continuously in a loop, whose execution can be ended by the user.

Figure 7.9 shows the LabVIEW block diagram for the VI which generates the control signals (VI 1 according to Figure 7.2). A number of three digital lines are accessed and set according to the requirements of the application. If we look at the shape of the basic control signals presented in Figure 7.5 we can identify the way in which they were created in LabVIEW. For example, the Pulse line, used for heating the capsule, contains two states: a High state lasting 14 ms and a Low state lasting 986 ms. If we look at Figure 7.9, the first setting in the Boolean Arrays indicates that the Port0 digital line is set to False for the first 14 ms (9 ms, first state +5 ms, second state) and then to True for the remaining 986 ms (981 ms, third state +5 ms, fourth state). Corroborated with the hardware implementation for the measurement circuit, the result is the Pulse control signal displayed on the oscilloscope screen presented in Figure 7.6. The obtained duration for the Pulse was of 14.4 ms which is sufficiently accurate for the proposed instrumentation. The RL_1 digital output line presents a period of 5 ms in which the analog input lines (AI0 and AI1) are

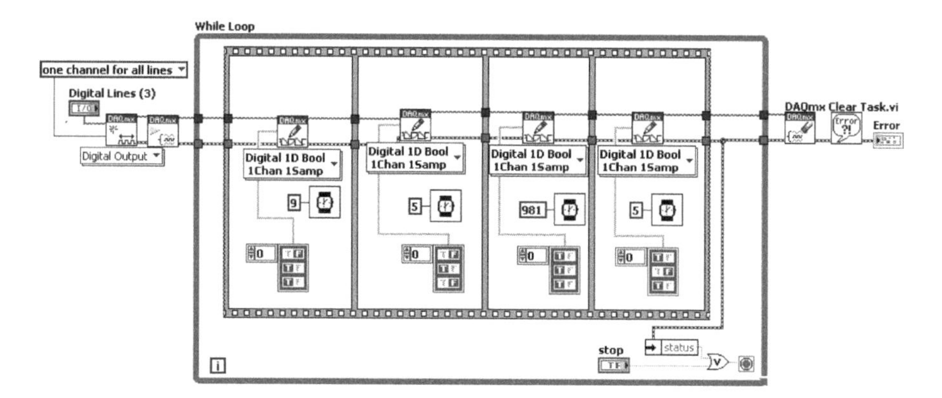

Fig. 7.9 Block diagram of the VI used for generating the control signals.

sampled. In Figure 7.9, the second row of the Boolean Arrays show that the digital line is set to False only in the last 5 ms. The rest of the time this line remains set to True. The oscilloscope testing confirms that the sampling of both analog input lines (indicated in Figure 7.8 as S-in and H-in) occurs in the second part of the 5 ms interval. That is between 997.5 ms and 999 ms when considering the diagram presented in Figure 7.5. The RL_2 digital line is set using the third row of the Boolean Arrays and is used in combination with RL_1 for applying the voltage across the sensor resistance.

Table 7.1 presents an overview of the control and acquisition lines and their connection on the CB-68LPR.

After the acquisition was triggered using a separate signal called DTV_1, the second virtual instrument executes a loop every 3 seconds. The results of this acquisition are two series of 1000 samples (one for each input line), one from the sensor's S-in line and one from the heater's H-in line. By examining recorded values from the series of samples, the program calculates the CO concentration and determines if the sensor is functioning correctly.

Regarding VI 2, indicators make up most of the front panel and display information on the sensor's output, functioning and previous measurements. Errors resulting from sensor element or heater malfunction are indicated. Data logging is performed if the user chooses to do so. If the sensor is not functioning correctly for more than 50 consecutive measurement cycles the errors are no longer logged. Measured data correction is automatically applied to measurement logged files. If a sampled sensor output voltage value is corrupted then it is replaced by a mean of the closest correct adjacent values.

Table 7.1. Overview of the connections on the CB-68LPR Table caption.

Connection no.	*CB-68LPR Pin*	*Signal Line*
1	14 (+5 V)	+5 V
2	52 (P0.0)	Pulse
3	17 (P0.1)	RL_1
4	49 (P0.2)	RL_2
5	69 (AI 0)	S–In
6	33 (AI 1)	H–In
7	64 (AI GND)	GND
8	67 (AI GND)	GND
9	47 (P0.3)	DTV1

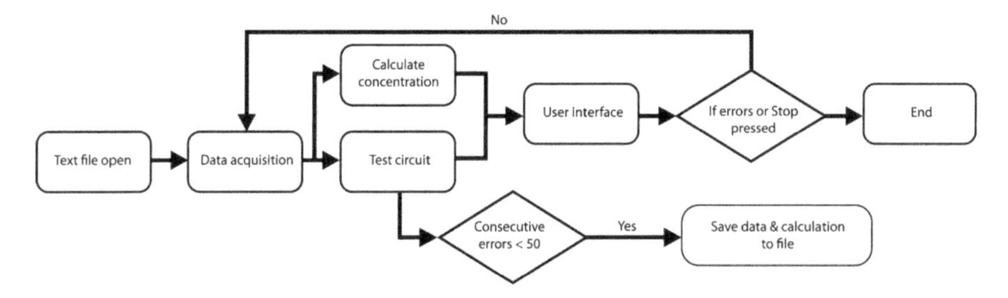

Fig. 7.10 Execution diagram for the second virtual instrument.

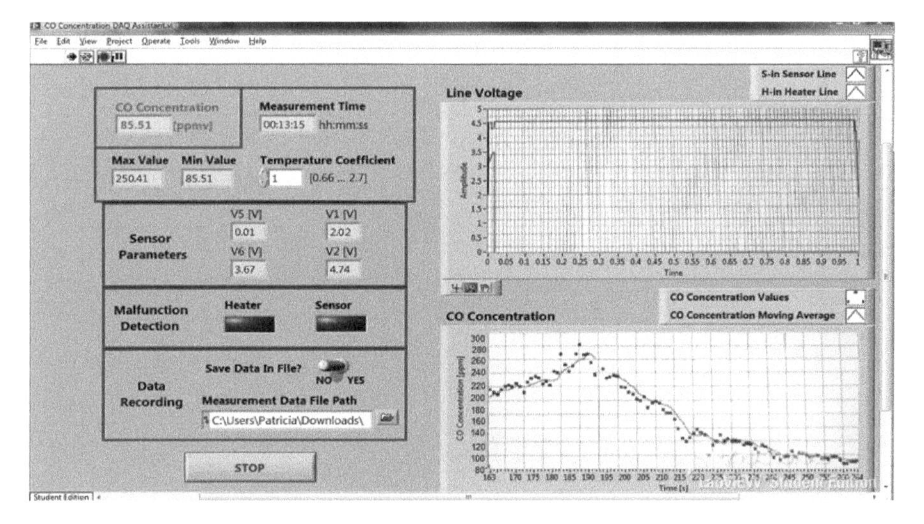

Fig. 7.11 The user interface for the CO concentration calculation instrument.

Figure 7.10 presents the execution diagram for VI 2. Figure 7.11 presents the front panel of the application.

The user front panel contains two graphs. On the Line Voltage display, the user can check the precision of the measurement operation signals. If the user chooses to verify how the Pulse line behaves, a supplementary hardware connection on the USB 6251 between the Port 0 digital line and a digital input line, provides the desired result. In Figure 7.11, the Line Voltage graph presents the voltage values acquired on the S-In and H-In lines in the previous measurement cycle. The CO Concentration graph presents the calculated values (with dots) and a running average over the last 5 measurements. On the left side, the indicators are used for presenting the concentration value in ppm, the extreme limits of the concentration values and the measurement

Table 7.2. Voltages for sensor capsule operation.

Sensor Element	*Correct*	*Malfunction*
Sensor	$V_2 > 0.5\,V$	$V_2 < 0.5\,V$
Heater	$V_5 < 0.1\,V$	$V_5 > 0.1\,V$
Heater	$V_6 > 3.5\,V$	$V_6 < 3.5\,V$

time of the last program execution. For the Max Value parameter, one can notice a value of about 250 ppm on the indicator while the CO Concentration graph presents a maximum value of about 290 ppm. The explanation resides in the fact that the graph holds the values for previous program executions (the maximum value of about 290 ppm) while the indicator is refreshed with each new execution. The Temperature Coefficient was set to 1. The sensor condition parameters are displayed both as numeric and LED indicators. Table 7.2 shows the parameters of the sensor capsule in the case of correct operation and when a malfunction occurs. By comparing the values displayed in the user interface and the values presented in the table, one can observe if there is a problem with the measurement circuit or with the sensor.

The front panel provides an option for data saving, if the user chooses to do this. For this particular application, the recorded data are presented as text files. These files include the date and time of the measurements, the voltage value corresponding to the sensor output and the calculated concentrations. If an error occurs during the acquisition process, the situation is reported along with the input line. For example, a heater problem will be indicated by the text "Heater Malfunction" and by the values from the lines V_5 and V_6.

7.6 Statistical Data Stationarity Testing

Because the voltage provided by the sensor's output has fluctuating values, the CO concentration can be correctly calculated only after the sensor has stabilized. Rigorous investigation of sampled values must be performed in order to indicate the moment when they are stationary. This implies that sampled data values will not vary significantly as measurement time passes. Only from this moment the correct CO concentration values are displayed [12, 13].

The application provides an automatic LabVIEW implementation of stationarity testing (using both a Run Test and a Trend Test) which can be applied for logged calculated CO concentration values.

Stationarity is tested for sequences of 100 concentration values which cover a time span of 5 minutes of data acquisition. For these segments of the random variable (calculated concentration for this particular case), 20 mean values and 20 mean square values are calculated. The median value of the data segments is taken as reference. The tests provide the possibility to calculate the degree (runs and reverse arrangements) with which the mean and mean square values fluctuate around the median. Final data evaluation values provided by these tests are compared against predefined values which are provided in reference [12].

In the case of the Run Test the values provided by the application express the number of runs over and under the calculated median value. This number of runs is a random variable r(k) with the mean value and variance expressed by relations (7.2) and (7.3).

$$\mu_r = \frac{2 \cdot N_1 \cdot N_2}{N} + 1 \tag{7.2}$$

$$\sigma_r^2 = \frac{2 \cdot N_1 \cdot N_2(2 \cdot N_1 \cdot N_2 - N)}{N^2(N - 1)} \tag{7.3}$$

The N_1 and N_2 parameters are the number of mean (and mean square) values which lie above and under the reference value. The N parameter is the number of values which are given at the input of the testing procedure ($N = 20$ in this case).

In the case of the Trend Test the values provided by the application express the number of reverse arrangements and are a random variable A(k) with the mean value and variance expressed by relations (7.4) and (7.5).

$$\mu_A = \frac{N(N - 1)}{4} \tag{7.4}$$

$$\sigma_A^2 = \frac{N(2 \cdot N + 5)(N - 1)}{72} \tag{7.5}$$

The LabVIEW implementation of the Trend Test is presented in Figure 7.12.

The Run Test block diagram is presented in Figure 7.13.

These algorithms were designed as subVIs and were included in a general VI called Test of Stationarity and presented in Figure 7.14. This routine was called in the case of recorded data for which additional processing was needed.

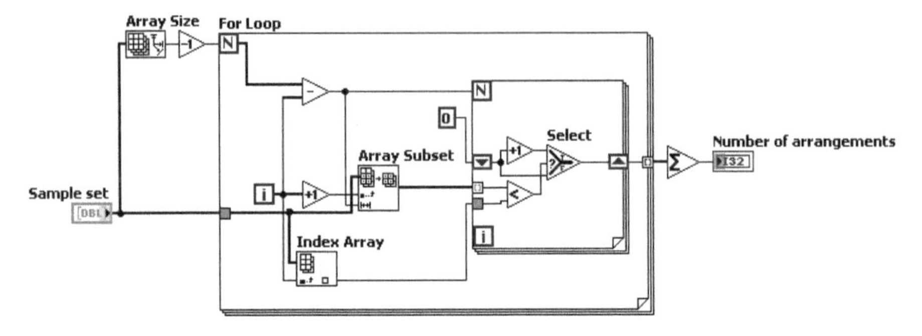

Fig. 7.12 Block diagram for the Trend Test.

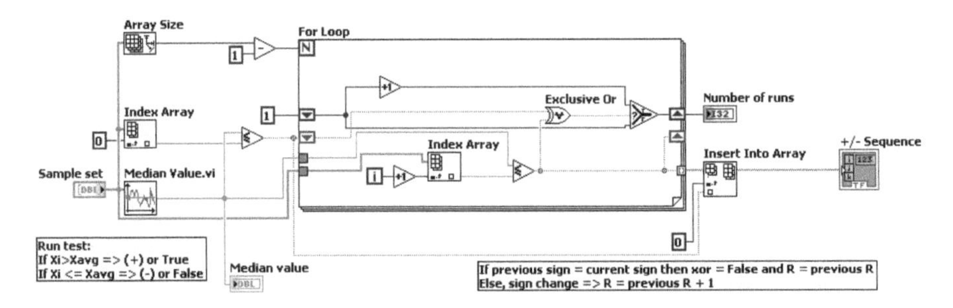

Fig. 7.13 Block diagram for the Run Test.

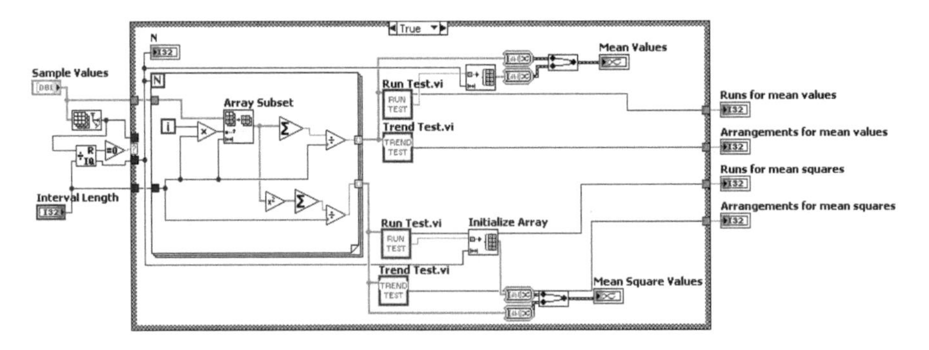

Fig. 7.14 Block diagram for the Test of Stationarity.

7.7 Experimental Results

The confirmation of the application's functionality was performed at an authorized car service company. In this way all measurements were compared with official, attested and legal results provided by the car service company. Some

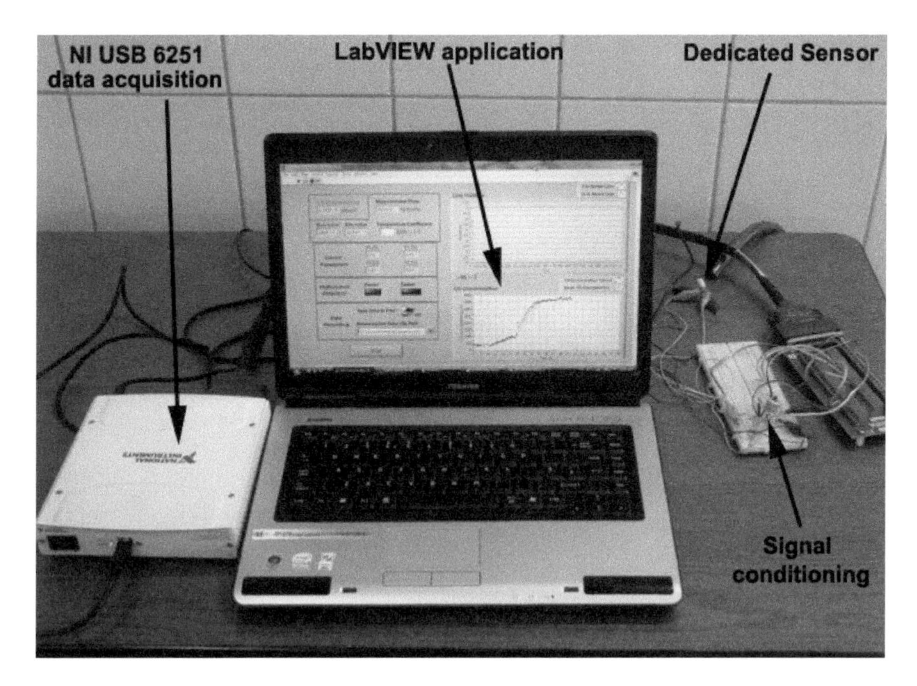

Fig. 7.15 Experimental setup.

results obtained for the case of a car with an engine running on petrol are presented below.

For exhaust gases parameters measurements, the car service company uses a Sun — MGA 1500 stand-alone testing device. The specialized device provides values for the CO_2, HC, O_2, CO, engine rotations and oil temperature [15]. The focus of our experiments was only the CO concentration values. Figure 7.15 presents the experimental setup. The elements of the proposed system are indicated with arrows and comply with the general concept of virtual instrumentation (Figure 7.1).

For the car on which the tests were run, the indicated value of interest was 0.01% CO concentration. Unaccepted levels for CO concentration start from 0.30%, if the acceleration pedal is not pressed or 0.20% with the acceleration pedal pressed. These values refer to an engine with a Euro 4 catalyzer. The tests which were performed with the virtual instrumentation application covered the case of the engine running without the acceleration pedal being pressed.

As the application was running, the sensor's output CO concentration value began a steady rise. This was expected since it requires a settling time of several

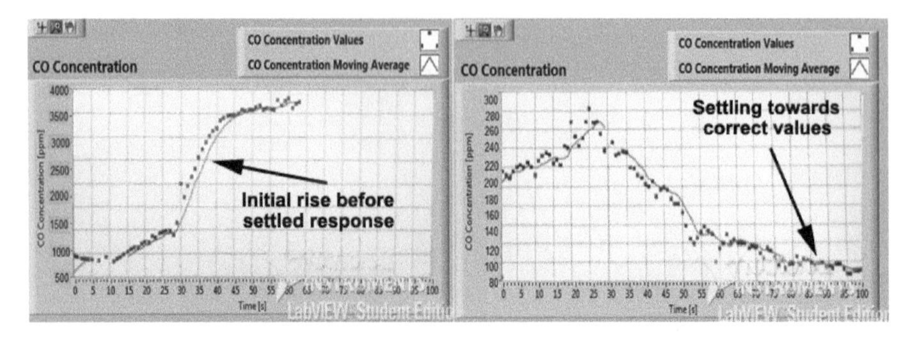

Fig. 7.16 Initial rise and settling of CO concentration values.

minutes. Figure 7.16 includes the initial rise and settling of the CO concentration time history. On the left side, a graph which presents the first use of the measurement system proves that such an evolution of the calculated concentration values is not valid. Values of over 3500 ppm are not valid for this particular experimental setup. This is the reason for which the measurement process was interrupted but the circuitry was not powered off. After several minutes, the measurement process was restarted and the calculated CO concentration values started to enter into a constant trend with limits ranging from 90 ppm to 100 ppm (the graph on the right side). These values indicate a concentration level with the lower limit of 0.009% CO concentration. The first two decimals indicated the same values (several measurements were taken) as the Sun — MGA 1500 dedicated device. In all cases both the sensor and the heater tests indicated no sign of malfunctioning. Voltages and calculated concentrations were logged in order to use them for further stationarity analysis.

7.8 Conclusions

In this chapter, a virtual instrumentation solution for measuring and calculating the CO concentration from car engines exhaust gases was presented. Using LabVIEW together with a solid state CO gas sensor in order to provide concentration measurements stands as an original concept.

The application uses the TGS 2442 sensor, the NI USB-6251 data acquisition device and the National Instruments' LabVIEW development environment. A dedicated circuit for the sensor was implemented and was controlled by a series of digital signals. In order to verify the proper functionality of the TGS 2442 sensor, three voltage values were sampled from the dedicated circuit. Another voltage value was sampled and used for CO concentration calculations according to the appropriate formula. Stationarity testing was implemented in order to provide a statistical tool for the analysis of the trends in the sensors output data evolution.

Results provided by the application were compared to the results provided by a stand-alone professional equipment. Insignificant differences between the results obtained by classic and the proposed instrumentation were attested. This determines the conclusion that the proposed instrumentation has been properly designed and implemented. However, one notices that the use of this particular virtual instrumentation means that there will be several minutes of incorrect measurements in the beginning. This fact is due to the settling time of the TGS 2442 sensor.

As potential application improvement the authors have identified two research directions. The first activity involves measurement of CO concentrations in controlled environments and a comparison with another dedicated device (the Testo 350 system). The second activity implies the integration of several gas sensors into a single virtual instrumentation solution and the transmission of the measurement results on a user mobile phone.

Acknowledgement

This work was partially supported by the strategic grant POSDRU/89/1.5/S/57649, Project ID 57649 (PERFORM-ERA), co-financed by the European Social Fund — Investing in People, within the Sectoral Operational Programme Human Resources Development 2007–2013.

References

[1] Torán, F., Ramirez, D., Navarro, A., Casans, S., Pelegri, J., Espi, M., "Design of a virtual instrument for water quality monitoring across the Internet", *Sensors and Actuators B: Chemical*, vol. 76, no. 1–3, pp. 281–285, 2001.

[2] Tooley, M., *PC-based Instrumentation and control*, Newnes, Oxford, 1993.

[3] Fountain, T., *History of Instrumentation*, Instrumentation Reference Book (Third Edition), Elsevier, pp. 802–860, 2003.

[4] Smiesko, V., Kovac, K., "Virtual Instrumentation and distributed measurement systems", *Journal of Electrical Engineering*, vol. 55, no. 1-2, pp. 50–56, 2004.

[5] Clark, C.L., *LabVIEW. Digital Signal Processing and Digital Communications*, McGraw-Hill, 2005.

[6] Chou, J., *Hazardous Gas Monitors*, McGraw-Hill and SciTech Publishing, 1999.
[7] Laboratory for Fuel Analyses, Environmental Investigations and Pollutants Dispersion Web Site, http://www.mediu.ro.
[8] The National Air Quality Monitoring Network, National Agency for Environment Protection, Ministry of Environment, Romania Web Site, http://www.calitateaer.ro.
[9] Skubal, L.R., Vogt, M.C., "Detection of toxic gases using cermet sensors", *Proceedings of SPIE*, vol. 5586, pp. 45–53, 2004.
[10] Llobet, E., Vilanova, X., Correig, X., "Novel technique to identify hazardous gases/vapors based on transient response measurements of tin oxide gas sensors conductance", *Proceedings of SPIE — The International Society for Optical Engineering*, vol. 2504, pp. 559–566, 1995.
[11] Figaro Engineering Web Site, http://www.figarosensor.com.
[12] Bendat, J.S., Piersol, A.G., *Measurement and Analysis of Random Data*, John Wiley & Sons, 1966.
[13] Papoulis, A., *Probability, Random Variables and Stochastic Processes*, McGraw Hill, 1991.
[14] Ionel, I. Ionel S., Nicolae D., "Correlative comparison of two optoelectronic carbon monoxide measuring instruments", *Journal of Optoelectronics and Advanced Materials*, vol. 9, pp. 3541–3545, 2007.
[15] Sun Diagnostics Web Site, http://www.sun-diagnostics.com.
[16] Lascu, M., *Tehnici avansate de programare in LabVIEW*, Editura Politehnica, 2007.
[17] National Instruments Web Site, http://www.ni.com.
[18] Dong, L.D., Sik, L.D., "Environmental Gas Sensors", *IEEE Sensors Journal*, Vol. 1, pp. 214–217, 2001.

8

Automated Measurement System Based on Digital Inertial Sensors for the Study of Human Movement

Konstantakos Vasileios[1,*], Nikodelis Thomas[2,†], Kollias Iraklis[2,‡], and Laopoulos Theodore[1,§]

[1]*Electronics Lab, Physics Dept, Aristotle University of Thessaloniki, 54124, Greece;* **vkonstad@auth.gr; §laopoulos@physics.auth.gr*
[2]*Biomechanics Lab, Dept of Physical Education and Sport Science, Aristotle University of Thessaloniki, 54124, Greece; †nikmak@phed.auth.gr; ‡hkollias@phed.auth.gr*

Abstract

Advancements in MEMS (micro electromechanical systems) have produced IMU (Inertia Measurement Unit) sensors that can precisely measure motion parameters that were traditionally calculated as derivatives of displacement measured parameters. These sensors have the potential to cover the majority of human movements. An implementation of such a system that uses two pairs of 3D accelerometers and gyroscopes and determines the exact position of a joint, by measuring the relative position of the two adjoined segments, is presented in this chapter. Such a system can provide joint kinematic information overcoming the shortcomings of video based motion analysis systems. Gait data are presented.

Keywords: MEMS sensors, accellerometer, gyroscope, motion analysis, gait analysis, orthopedics.

Advanced Distributed Measuring Systems — Exhibits of Application, 183–210.

8.1 Introduction

Motion analysis is a process used to define the momentary position of the human body in space based on its particular land marks, with specific frequency and then calculate its kinematic and/or kinetic parameters. In conjunction with other synchronized devices, such as dynamometers or electromyography etc., it helps the researcher, the coach in sports, or the physician to evaluate movement and detect its disorders.

Traditionally, the basic tool for Motion Analysis is a series of consecutive photographs of movement recorded with predefined frequency. Then, during the data reduction process, the position coordinates of body landmarks are derived using analogies of predefined and measured points in the field of view of photographs.

From 1892 when E. Muybridge presented his work "Animal locomotion" (Figure 8.1) up to now there has been a great improvement of systems that record movement. It started with cinematography pictures; it went through high speed cinematography and then video pictures, arriving to the state-of-the-art optoelectronic systems (that automatically detect the position of active or passive markets located on body landmarks).

Since a photograph is a plane representation of a three dimensional space, researchers soon tried to figure on how they can get 3D coordinates from

Fig. 8.1 Animal locomotion.

2D pictures. The work of Abdel-Aziz & Karara in 1971 [1] is the foundation of three dimensional analysis of today's systems. They developed a photogrammetric method named DLT (Direct Linear Transformation) that can reconstruct space coordinates from image coordinates of at least two pictures of the same space, recorded from different positions. The use of a number of points in the field of view with well-known 3D coordinates, the so called "calibration tree", is necessary.

Modern systems use more than two synchronized cameras, and instead of calibration tree they use an object of well-defined geometry that moves in the field of view. Its consecutive recorded positions from all cameras give enough information to be used for the reconstruction of space coordinates.

A main step in the motion analysis is the digitization of consecutive pictures. Traditionally, this is done manually, by digitizing each point of interest for every frame for all cameras. This is a time consuming process that also includes a great risk of human error. In cases where the researcher cannot "touch" the subject for putting markers on his body, such as an athlete during competition, this is the only way of digitizing. However, in laboratory conditions subjects can be "loaded" with markers on specific body landmarks that can be used for this process. There are two kinds of markers, the so called "passive" and "active" markers. The first kind reflects light, usually IR, that comes from a system around each camera and its position is recorded, while the second kind produce light in order its position to be recorded from the cameras. With both markers, digitization of points is achieved automatically by scanning the recorded images.

After the reconstruction of the 3D coordinates of the body landmarks, kinematic and kinetic parameters of body joins, segments and total body, such as angles, velocities and accelerations are calculated by applying numerical methods to the position coordinates.

When measuring in nature, there is always an amount of noise or so called error inherent to measured data. The accuracy of measurements with any of the existing imaging system for motion analysis depends upon the analogy between image resolution of cameras and the dimension of recorded space, the shutter speed of cameras (time that takes to record each image), the accuracy of digitization, and the selected numerical methods. Although human motion is slow in nature compared to other physical phenomena, recording frequency is of great importance, especially when time derivatives of motion are required.

To choose a hardware system for motion analysis, one must compensate between image resolution, recording frequency, and cost. Usually, high resolution systems are low frequency, and vice versa, while a high resolution and high frequency system is a system of very high budget.

Even so, methodological limitations of the high tech — high budget systems are still present. Specially trained personnel is needed to operate them, while they cannot easily operate out of the lab, capture long duration movements, or movements evolving in a large space. Finally, there is always the time of preparation which is frustrating for the subjects, although this is a compensation for better quality of the data.

The so called "commercial systems" are based on low cost commercial video cameras with restricted recording frequency up to 50/60 Hz, but no other shortcomings. The accompanied software of such systems, given for recording of motion, digitization and data reduction, can produce accurate information for position and even displacement of low frequency movements, yet the calculated derivatives can be problematic [2].

Advancements in micro-electromechanical systems (MEMS) have produced IMU (Inertia Measurement Unit) sensors that can precisely measure motion parameters that traditionally were calculated as derivatives of displacement measured parameters. They are miniature in size with low power consumption and can measure 3D accelerations and angular velocities, as well as global orientation parameters when supplied with magnetic elements. These sensors in conjunction with high tech electronic circuits are still affordable in cost. They allow data collection during unconstrained continuous movement over prolonged periods of time and they have the potential to cover the majority of human movements. Although this technology sounds as a good solution, there are still challenges that have to be met. Such issues concern the extraction of movement-related information from the signal derived, which can be affected by offset errors that rapidly accumulate over time and sensor wide oscillations caused by the inertia of soft tissues of the human body were they are placed [3]. Issues like global referencing need to be handled with the use of magnetometers, drift of the gyroscopes must be compensated by the parallel use of accelerometers and magnetometers [4]. Most of the existing systems that do not have the subject wired to a PC, use data collection devices like flash memories or SD cards, that have to be attached to the subject's body thus data can't be processed in real-time.

Even more, the necessary hardware that the examinee has to carry on him is still heavy, not small enough in size and obstructs movements. As in other acquisition systems like EMG devices for example, the case of RF communication restricts movements near to the work station thus attenuates the advantage of portability of the system. Furthermore, at most of the reported works, noise during data collection affected the data that had to be filtered before further processing [5], which of course is a standard process used with every acquisition method [6]. Weight/mass, size, power consumption, placement on body landmarks, integration of synchronized multiple sensors solution for full body analysis, combination of video information in order to comprehend movement patterns through vision, proper software for error free data, validation of the methods and the models used, discriminate the present and future research ground.

The present status of the research ground is illustrated in a representative review [7] which is related to the measurement of human lower limb movements with the use of inertial sensors. A systematic search in several search engines (ISI Web of Knowledge, Medline, SportDiscus, IEEE Xplore) indicated a large amount of papers (39) with related topics. Several methodological issues were explored there like the type of sensors being used, the fixation methods, the data logging algorithms and the processing stages, in several case studies and applications. A brief summary is presented on Table 8.1, which is partially adopted from the original review.

It is worth mentioning that although video/optical systems are widely used and papers are published validating their methods just by referencing the system they use, there are methodological studies that identify the limitations of these analyses and try to mathematically optimize their methodological steps in order to reduce the error and produce more valid data [8]. One of the problematic issues of the video/optical systems is that high resolution and high sampling rates, which are important features for good data quality, are very expensive.

Before trying to validate sensors data with the parallel use of video/optical based motion systems it is important for one to comprehend the fundamental differences of the data processing methods of image based motion analysis and IMU based methods.

First of all it should be made clear that every method of measurement is subjected to errors. Apart from the random (white) noise during data acquisition,

Table 8.1. Range of the methodological parameters as they appear on the literature.

Sensors type, ranges	• accelerometers, gyroscopes, magnetic sensors (uniaxial, triaxial) • range: accelerometers 3 g to 10 g, gyroscopes 300 to 1200°/s, magnetic sensors: 750 m Gauss • weight: 18.2 g to 700 g, size: $20 \times 10 \times 7.2\,mm^3$ to $64 \times 62 \times 26\,mm^3$ • Sampling frequencies: 20 Hz to 800 Hz
Logging Type	• Portable data logger (SD Cards, flash memories), hand held PC • Wired notebook, Bluetooth communication
Processing	• Low pass filters (cut off frequencies 15–40 Hz) • Filter types: Butterworth, Kalman, Savitzky-Golay • Curved fitting techniques
Study Design, Validation	• Young recruits (age 18–40), older subjects (aver. age 58.7) • Sample size ranged from 1 to 36. • Cases: Walking and running, stand-sit transition, landing, tennis serve, rowing, cycling, jumping, knee and ankle joint movement. • Comparison with video cameras or high speed optical motion analysis systems with reflexive markers.
Applications	• Joint kinematics of ankles, knees, hips (simplified, detailed 3D) • Tibial acceleration • Case studies: Analysis of skill level and locomotor performance of athletes or patients, ambulatory measurement to monitor patients' daily activities, clinical assessment for patients, gait event detection and analysis, identification of different daily activities
Fixation Methods	• Velcro straps, double-side adhesive tape, elastic straps, neoprene straps, sensors fixed on aluminium plate or put inside plastic casing, semi-rigid belt, exoskeleton harness • Anatomical calibration to align sensors' axes with the axes of the body segment (calibration devices, static postures)

which is attributed to many factors, there is also an error that data inherent during processing. A distinction should be made at this point between video data and data from IMUs and especially accelerometers and gyroscopes.

The video based systems record displacement. To begin with, the resolution of these quantities are subjected to the combination of time errors, due to reduced refresh rate and to space errors, due to the fact that some key events, like the exact moment that a foot touches the ground in gait analysis, are not captured precisely. For example, the touchdown happens between two frames, thus no actual exists with the exact moment. The general rule for handling measurement errors on video based motion analysis of human movement is to filter the displacement data with a low pass filter, no more than 5–7 Hz cut off frequency, depending on the movement under study [9]. Furthermore the mathematical action that takes place in order to create velocities and accelerations are multiple differentiations. If a differentiation is to be calculated numerically, then it is described by the expression

$$f'(x) = \frac{f(x+h) - f(x)}{h} \tag{8.1}$$

The error in such a mathematical action is described by equation

$$R = \frac{f(x+h) - f(x)}{h} - f'(x) \tag{8.2}$$

If Taylor's Theorem is used, then the error is calculated as

$$R = f''(x)h/2 + f'''(x)h^2/3 + \cdots \tag{8.3}$$

If the first factor is considered as more significant, then the produced error is related to f" and the quantity h. So, the error depends on the sampling period, and the 2nd order derivative, which is related on the stability of the signal. This factor actually expresses the rate of change of the acceleration.

On the other hand, the IMU sensors record the quantity of linear acceleration or rotating velocity and one must calculate integrals in order to produce eventually displacement. These measured quantities suffer fewer errors, because they combine better resolution and increased sampling frequency. Those errors mostly concern white noise, temperature drift and offset. Their handling includes mostly temperature compensation and filtering. A common applied filter is a Kalman filter, which is a set of mathematical equations that provide efficient computational (recursive) means to estimate the state of a process, in a way that minimizes the mean of the squared error. In this case, it uses measurements observed over time, containing noise and other inaccuracies, and extracts new calculated values according to the aforementioned criteria. In a similar numerical way, velocity is calculated out of acceleration as [10]

$$\begin{aligned} u_V &= \sum_{i=1}^{v} \tau v_i = \tau \tau \gamma_1 + \tau(\gamma_1 + \gamma_2) + \cdots + \tau(\gamma_1 + \gamma_2 + \cdots + \gamma_v) \\ &= \tau^2 \sum_{i=1}^{v} (v + 1 - i)\gamma_i \end{aligned} \tag{8.4}$$

The error in such an expression is

$$\sigma_{\upsilon} \approx 0.6\mathrm{T}^{2}\sigma_{\gamma} \tag{8.5}$$

The error is related to the total duration of the recording and the measured error of acceleration.

It is apparent that both of the data processing methods involve error, either on the measured data, or on the mathematical processing (derivation and integration). The video analysis is more prone to errors due to the decreased resolution of the measured data and the mathematical processing afterwards that depends on variables that come with these errors. On the other hand, either the measured data or the mathematical processing in the case of the IMU sensors seem to be less influenced by those factors. Nevertheless they are also prone to other sources of error like for example fixation errors caused by skin movement or by improper alignment of the sensor's axes to the anatomical axes of the segment.

Overall, IMU sensors have the potential to be an important tool in the area of kinematic analysis. In fact, there are already companies, such as XSens Technologies, that provide motion analysis packages based on such sensors. Apart from that, accelerometers, gyroscopes and magnetic sensors are available in the market for whomever has the know how to build a system according to his needs.

In the following text a custom implementation of such a system is presented in detail, describing the specifications, the selected hardware and software algorithms, and the necessary data processing, in order to be applicable in general purpose two segments — one joint 3D kinematic analysis. The use of this implemented system is presented through a case study example of a gait analysis application.

8.2 The Instrumentation System

8.2.1 Specifications

The system is built based on the needs of the study of human movement. Specific restrictions/demands apply to such a research area. First of all, the system must be small in size/weight and highly portable. The human subject should not feel uncomfortable or be restricted in moving. The experimental conditions must not be limited by the lack of portability. These facts create the necessity of low power implementations, reduction of the battery size and increased time duration that the system can abstain from a power supply, thus obtain better portability.

An arising issue is the exact measurement of a moving segment in 3D space. This requires at least three degrees for translational and three degrees for rotational quantities, providing that they are related to a global referencing system. Alternatively, at least in the case of a two segments movement, even the local reference system of each segment can be adequate to describe movement in a comprehendible manner.

When working with markers one would have to use at least three markers for each segment, which should not be collinear and be placed in adequate distance, in order to form a level which would rotate and translate. Those markers are usually placed in anatomical body landmarks, like condyles.

At the sensors solution each IMU has its own reference in the sense that every consecutive sample can be related to its' previous. For the specific implementation that measures two adjoined segments, a 3D gyroscope and a 3D accelerometer must be used for each segment. The two sensors must have one of their axes aligned to the longitudinal axis of the segment and be placed in parallel to each other. The accelerometer measures the linear kinematic quantities and the gyroscope the rotating ones. One set of sensors for each segment is adequate. Every parameter that is needed for the analysis, like joint angle, can be calculated by combining each sensor's information using the proper mathematical formulas.

The rate of the measurements is another important thing that should be taken into consideration. Taken the fact that the sampling rate must satisfy at least the Nyquist-Shannon sampling theorem, there are movements (like long jump in sports) that need to be sampled with 200–300 Hz. Even at gait kinematics, that the spectral content does not exceed 5 Hz [11], sampling rates like those of the low cost commercial video cameras (25/30 Hz) can be considered as an acceptable solution only for the calculation of spatiotemporal stride parameters, as well as angular and linear displacements. When it comes to joint torques and joint powers there is a loss of information through the derivation process, when comparing for example the 25 Hz with the 100 Hz sampling frequency. This phenomenon has a homologous effect with over smoothing [2]. On the other hand up-sampling the signal to higher frequencies through interpolation is a tricky task and can lead to erroneous data.

MEMS sensors technology has no restrictions at such matters. For instance, sampling rates of the sensors are more than adequate, the signal is digital, while the fact that gyroscopes and accelerometers measure angular velocity and linear acceleration may give an advantage in the precision when calculating inverse dynamics.

For the presented system sampling frequency is located between values from 50 Hz to a couple of kHz. As it is for told, 50 Hz is an adequate sampling rate for most common human movements, with 200 Hz being a more than adequate rate for almost every human movement, except situations where collisions occur, in which a rate of a couple of kHz is required.

The type of data transfer for the IMU solution, real-time or other, is already addressed as an issue that can compromise the quality of the measurement as it deteriorates it in many ways and obstructs the subject's performance.

Two alternatives are implemented in the present case. One implementation is with a memory module where no wires are needed. This can enhance the portability of the system. There is also the real-time way, where wires and instant previewing is needed and can be used for small scale movement study, where the previewing is more important.

The specific implemented system is based on the usage of a central microprocessing unit that is in charge of any possible functionality and module handling. A modern microcontroller is more than enough in such a case, with adequate processing power, communication functionality, and the ability to reprogram for easy development.

Finally, there is always the need for a convenient way of controlling, configuring and exchanging data with a personal computer. For this reason, the graphical programming suite LabVIEW is selected, as it provides advanced functionality, with a friendly interface, in both the development and the final operating stage.

8.2.2 Hardware Description

The block diagram of the implemented measuring system is illustrated in Figure 8.2. It consists mainly of three parts: the sensors, the processing and storage part, and the communication modules [12].

The main module of the system is the microprocessing unit; a modern typical 8-bit low power microcontroller, with adequate functionality to implement

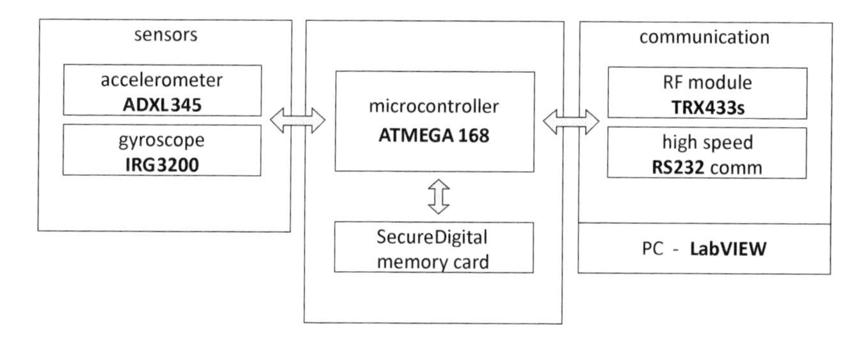

Fig. 8.2 The proposed implementation.

the measuring procedure. The microcontroller chosen is ATMEGA168 from Atmel. It operates with an internal clock of 8 MHz, it has adequate number of general input/output pins, 32 Kbytes flash memory, 512 bytes EEPROM, 1024 bytes RAM, and it supports several serial communication interfaces.

The storage component is a common Secure Digital (SD) memory module of 2 GB capacity. These memory modules are quite common on the market, at a real low price, with reading/writing speed adequate for low/middle acquisition speed applications. The operation of such a module is quite easy when combined with a microcontroller, by utilizing a typical synchronous serial peripheral interface (SPI). The writing procedure in this module supports data packets of 512 bytes, and not only single 1 byte actions. So, a write access occurs when 512 bytes of data are already obtained. The time needed for a 512 bytes write is not negligible, and varies in the scale of a few milliseconds. This means that the system must be smart and fast enough so that while writing to memory, no data are being lost. This delay is not of a constant value and may vary when writing to different memory modules, or even when writing to different locations of the same memory module.

The source of data is presented on the left part of Figure 8.2, and consists of two sensors, a digital accelerometer and a digital gyroscope. The accelerometer provides 3-axis values in every sample, which is one 13-bit number for each axis. Each value corresponds to an acceleration vector component, related to each axis of the Cartesian system of the accelerometer. The chosen module is ADXL345 from Analog Devices. This device can provide a rate up to 3200 samples per second, at a supported bandwidth of 1600 Hz. It can measure accelerations to a maximum range of ± 16 g. Many of these parameters can be assigned to different values, according to the specification of different applications, and more specifically to quite lower values, since human movement is generally characterized by small frequencies. It can communicate with all common serial protocols (SPI, I^2C). The direct digital nature of the resulting data is vital, since it minimizes the usage of external modules, like A/D converters, while on the same time the provided accuracy is much better since the complete accelerometer chip is created with high quality standards.

The second sensor is the gyroscope ITG-3200 from InvenSense. This device also provides data in digital form. Each sample measurement consists of 3 values, the rotating angle velocities along each one of the axis of the 3-axes Cartesian system of the device itself. Each sample is a 16-bit number (15 bit + 1 bit sign). It can measure angle velocities up to a maximum range of $\pm 2000°$/sec. Also, it can provide a rate up to 8000 samples per second. Again, some of these parameters can be configured at more relaxed values, according to the specifications of each application. It supports the I^2C serial communication protocol. The digital nature of the provided data is also an important advantage for the reasons mentioned on the accelerometer. Furthermore, an internal programmable low-pass filter exists within the chip for filtering the measuring data according to each application's needs.

Also, communication modules appear on Figure 8.2 for being able to interact and exchange data with the outside world of the instrumentation system. One of these modules is the RF module TRX433s by RF Solutions. This is an FM Transceiver operating at the frequency of 434 MHz. It can Transmit/Receive at a rate up to 115 Kbps at a maximum range of 300 meters. This module is used for receiving start/stop recording signals from a remote control. Such signals mark the measuring time window, so that the system does not record all the time, but only when needed. Also, these signals can place time marks on the measuring procedure, so that specific events can be easily spotted on the resulting data. In future development, this module will also be responsible for transferring wirelessly the recorded data to a computer.

A wired communication module also exists, a Serial-to-USB converter, for transferring the recorded data to a personal computer, and to facilitate real-time applications. The device implementing this conversion is the cable/converter TTL-232R-3V3 from FTDI, which includes the necessary electronic circuit, physically close to the USB jack, for the conduction of the conversion procedure. A convenient speed of 500 kbits per second is supported, which is considered quite a high speed for serial communication between data acquisition systems and a personal computer (approximately 50 kb/s).

As mentioned already, the serial communication is being held with the microcontroller and a personal computer. The software running on the computer is a user's interface implementation within the LabVIEW graphical programming environment. LabVIEW is characterized by increased functionality for fast and rather easy creation of instrumentation systems. It has a large number of functions related to serial communication and string handling, so the exchange of data with the measuring system is made quite easy. It also provides multiple ways of plotting data without the need of time consuming programming, and it gives an adequate number of file handling functions, so that the resulting data can be stored on disk, and furthermore opened by other software for further processing.

8.2.3 Instrumentation Procedure

Up to this point, the complete measuring system is presented in terms of hardware choices. Modern modules have been selected, with adequate characteristics for the needs of the application under study. The next step is the development of the necessary software routines, either on the microcontroller or on the computer. This programming part is of significant value, since the advanced possibilities of the chosen hardware are best exploited when they are combined with high quality programming/instrumentation procedures. Figure 8.3 illustrates the block diagram of the measuring procedure, as it appears on the code that runs on the microcontroller.

Initially, the measuring system is having a reset, immediately after getting power supply. The power supply is a battery of a value of approximately 3.7 V. The reset is followed by the initialization of the various components of the measuring system. The sensors are being configured according to the necessary speed and accuracy specifications, the memory card is being prepared for proper use, and the RF module is configured for accepting start/stop signals.

After the initialization procedure, the system stands idle, by placing several modules to a medium level of low power state. This state will be interrupted either by an incoming start/stop signal or a serial communication action from the computer. The response of the microcontroller to the component modules is interrupt based, for low power purposes and for creating fast responses to

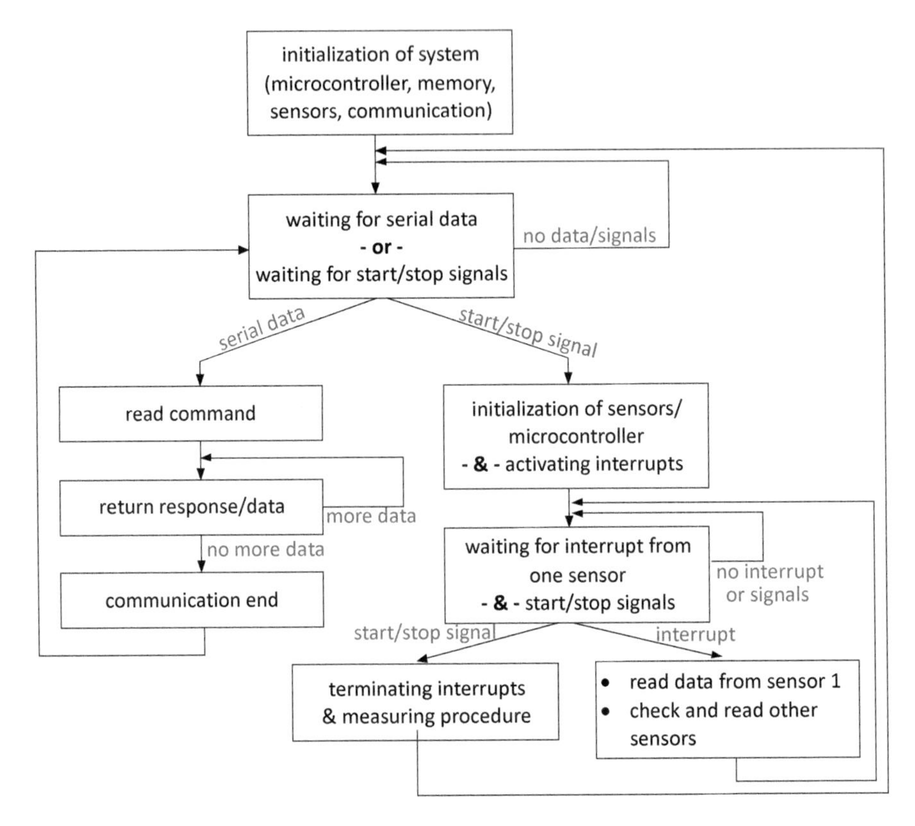

Fig. 8.3 The measuring procedure.

the requests of these modules. So, the RF module is listening for wireless requests and informs the microcontroller immediately if a request is received. Also, a serial communication may occur. When a request is available, the microcontroller responds accordingly.

If a start/stop signal is received, it initializes the necessary variables for the measuring procedure, and begins the measuring algorithm. During this algorithm, first the microcontroller waits until one of the accelerometers has available data, according to the already configured speed and accuracy settings. The accelerometer will provide an interrupt signal when data are available. Upon the receipt of this signal, the microcontroller requests the data from this accelerometer. After that, it requests data for the other three sensors. For every sensor it firstly does an initial check that data are ready. When the four sensors are read, the system returns to the waiting state for the interrupt signal from the first sensor, and this acquisition procedure is being repeated, until a new start/stop signal is received.

The acquired data is not written immediately on the SD memory module. As already mentioned, data must be written on this module in groups of 512 bytes. So, the measurement data are being stored temporarily on a buffer on the memory of the microcontroller, and when 512 bytes are gathered then they are sent on the memory module. Also, the SD writing procedure is quite slow compared to the rate that new data are being gathered. Thus, while the writing procedure on the SD card lasts, the measuring system is still collecting data, but with a temporary storage on an alternate memory buffer on the microcontroller. The result is two alternative buffers, one collecting data and the other sending its stored data to the memory card. Even, with this collecting technique, the memory card must be fast enough so that the writing procedure is finished faster than the time needed for filling the alternate buffer. Not all SD memory cards are fast enough for such a procedure, so a proper SD card research is needed when hardware selection is occurring.

This measuring algorithm is being executed repeatedly, until a new start/stop signal is received from the wireless module. If a new start/stop signal is received then the microcontroller terminates the interrupts related to the sensors, and turns again to the initial low power idle state, for accepting new data from the wireless module, or the serial port.

If incoming serial data appear, the microcontroller responds accordingly by either configuring parameters of the instrumentation procedure, and by returning data to the computer if requested.

This is the normal acquisition mode of the system, where data are being stored to the SD card. But the system can be configured, by the interface running on LabVIEW, to send the acquired data immediately to the computer, thus converting this system to a wired but real-time acquisition system. This alternative mode is described by the same measuring algorithm, with the only difference being the part where the system stores data to memory. Besides that, every other operation is the same.

The data in either real-time or no real-time mode are accepted by the same interface on LabVIEW. These data are a combination of 30 bytes for each sample (24 bytes from sensors, 2 bytes from temperature, 1 byte for timing purposes and 3 other bytes for communication control). Upon receipt, these data arc converted to acceleration and velocity, and temperature compensation is applied. When non-real-time data are received, they are immediately stored to a file in a raw form. Then, they can be previewed and exported at any time. When real-time data are received, the procedure is also the same, but with advanced attention to the previewing of data, because this is the major need in this situation.

Further processing of the data can occur within the LabVIEW suite, but also on other software, e.g., Excel and Matlab. LabVIEW is really convenient for the first steps of data handling, but it becomes complex when more advanced processing is required or when a large amount of data is to be processed. So, simple integrals of the measured linear acceleration and rotating velocity are calculated and previewed on LabVIEW, but at the end Matlab becomes the main means of processing.

The processing that must be made is related to the parameters of the human motion that are to be studied. In the case of gait analysis, spatiotemporal characteristics and angle histories during a gait cycle is of primary importance. Therefore an integration of angular velocities must take place to calculate the angles, while the spatiotemporal parameters like stride time and length, percentage of swing and stance phase in the gait cycle can be derived from the accelerometer measurements.

8.3 Performance Characteristics

The description of the system under study indicated the hardware that is used and the algorithms (software) that are implemented. The result of this software-hardware combination is a complete measuring system with some specific performance characteristics that rely on parameters based on the hardware selection, the software implementation and the combination of these two.

8.3.1 Accuracy, Resolution

The measuring accuracy is an important part. Possible sources of inaccuracies of the developed system are the digital step of each measurement, the noise that is induced in every waveform, data sampling period fluctuations due to clock fluctuations of the microcontroller, offset drifts of the sensors acquired data due to temperature variations, and finally inaccuracies produced by the mathematical calculations, e.g., from the integrals that produce velocity and displacement.

First of all, the digital step is related to the A/D resolution that operates on the sensors. The accelerometer uses 13 bits to describe a signal of full scale ±16 g. This leads to a digital step of 4 mg/LSB. On the other hand, the gyroscope uses a 16 bit resolution A/D converter to describe signals up to ±2000°/sec, leading to a digital step of 0, 07°/sec (or 14,375 LSB per °/sec as described on the datasheet). Both digital steps are quite small for the indicated research area, and especially in comparison with tools that already exist.

Also, noise exists in the measurements of both sensors. The accelerometer data fluctuate approximately 0,1 g due to noise, and the gyroscope data approximately 1°/sec. These values are not completely negligible and proper handling should be considered when calculations are being made e.g., average values.

As far as the sampling rate is concerned, possible inaccuracies arise also from there. This sampling rate depends on the operating clock of the microcontroller, which means dependence on an oscillator that feeds the microcontroller with a specific frequency e.g., 8 MHz. Possible fluctuations to this oscillating frequency are a cause of inaccuracy because of the mathematical action of the integration (multiple integrals in case that linear displacement is needed) that must take action later on. Nevertheless, these fluctuations that are of the scale of below 1% are negligible in this research area, due to the low sampling rate that is needed when the human motion is being recorded.

Inaccuracies become important when temperature drifts exist. Both sensors, and especially the gyroscope, have a temperature coefficient that should be taken into consideration. For such a reason, the gyroscope provides also a temperature value, along with the sampled rotating velocity. The measuring algorithm should read this value and compensate the measuring data accordingly. This procedure provides data that are almost independent of temperature changes. Since the related temperature range is close to room temperature, a simple linear compensation is considered adequate. If a larger range is to be considered then probably a more complex equation is required.

The error propagation in such an application must be taken also into consideration. As described already, the needed mathematical action is the integration of the measured data, so that linear velocity and linear and rotating displacement are calculated. The errors that appear in these mathematical actions should be calculated by the usage of the proper error propagation analysis.

8.3.2 Sampling Rate

As it is described in the sensors specifications, the accelerometer can measure up to 1600 samples per second, while the gyroscope can measure up to 8000 samples per second. These values are more than adequate for the description of the human motion. Therefore, they create quite relaxed operating conditions for the sensors, for the instrumentation procedure implemented on the microcontroller and for the demand of speedy SD memory cards.

8.3.3 Portability

One of the most significant specification constraints is the size of such a system. There are existing systems based on MEMS technology that can easily be used on the field since they can be implemented in small, portable and, in terms of consumption, low power solutions that are not constrained from stationary units, such as receivers and cameras [13]. Most of these features are important for the subject under study, since he must not feel discomfort or uneasiness when the system is putted on. The specific implemented system is quite small in size (approximately $80 \times 50 \times 30\,\text{mm}^3$ for the main system and $30 \times 20 \times 20\,\text{mm}^3$ for each sensor module). Its' use does not restrain the subject in any way, thus making the movement that is being recorded more natural. In video analysis systems, the need of wearing markers creates quite the opposite effect for the subject. Even more, small portable systems like that one (Figure 8.4) are ideal to be worn outside a laboratory. Each measurement can be made for a significant larger amount of time, since the subject is free to move without constrains, in contrast with a few seconds movement that video systems are able to handle. Gait analysis is a good example. Especially in cases of clinical gait analysis those characteristics are determinative, since a patient with a disability usually prefers to perform at his natural environment and not to visit a laboratory. This attitude can have a psychological effect on his performance during the measurement.

8.3.4 Energy Consumption

The energy characterization is significant information for such a system. These kinds of systems operate with a power supply based on a battery, with low power operation being one of the main concerns on the development procedure.

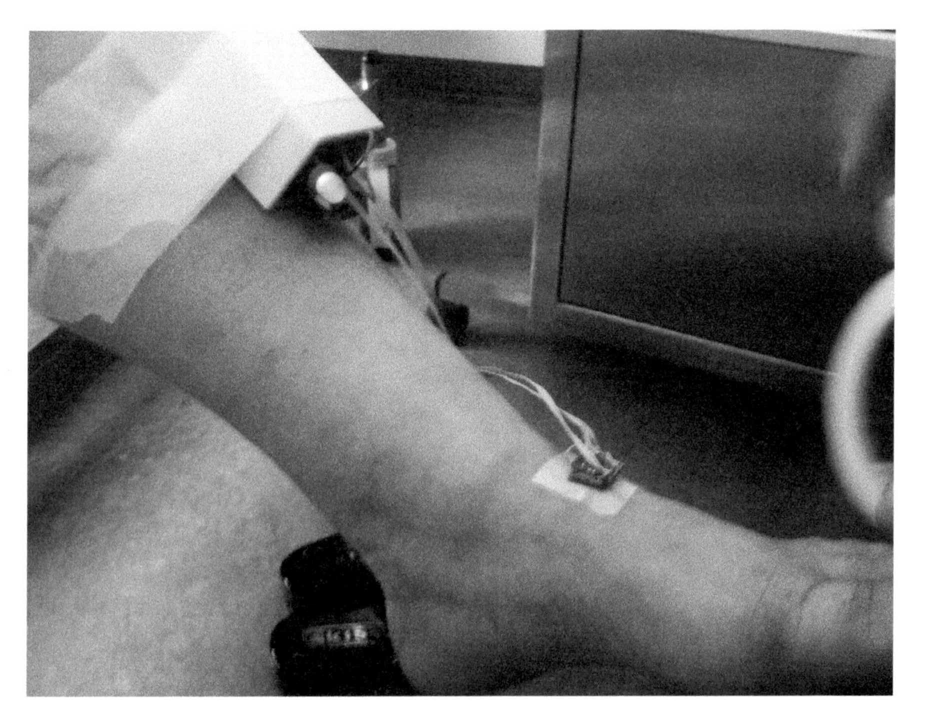

Fig. 8.4 Photo of the system placed on a person.

The desired directive is to continuously make more powerful systems in terms of functionality, but less power consuming in terms of energy. On the system under study, a modern microcontroller, modern sensor systems, and more or less all modern available modules, consume already less in comparison with prior modules and also support alternative means of operation with reduced consumption (low power modes). Special effort must be made on the development procedure so that the energy behaviour to be recorded and eventually improved. This recording of energy data can be made with simple means, like an ammeter, or with more complex means, like the implementation of [14], where specific functions implemented on the modules of the system can be assigned with energy measurements, thus creating a more detailed energy characterization.

In practice the microcontroller puts itself and some of the modules in low power mode when needed. As already mentioned, the microcontroller operates in an interrupt driven way. While not being in an interrupt it goes to an idle state with reduced consumption (average current $\approx$ 3 mA), and also leads the sensors to go to a low power mode (average current for all sensors $\approx$ 1 mA). The rest of the modules are left untouched, since the RF transceiver (average current $\approx$ 10 mA) needs to be fully operating all the time, so as to accept signals at any time, and the SD card is not consuming any current since all memory modules of flash nature usually consume current dynamically, when they are being used (read/written) and not statically, when sitting idle. Out of the result of current measurements, the system appears to consume 14 mA while standing idle and 35 mA in average values when operating in acquisition mode. If a battery of 3.7 V and capacity 900 mAh is to be used, this provides the ability to operate several hours either on standby or acquisition mode.

8.3.5 Functionality — Automation

The combination of a microcontroller and a personal computer provides a large scale of functionality. First, the microcontroller is a small-sized microprocessor. It is able to communicate with quite a large number of modules with various serial interfaces. In this specific implementation, 4 sensors, 1 RF transceiver, 1 SD card and 1 usb-to-serial module are being used. The microcontroller can utilize these modules in the best possible way by proper configuration. It is also convenient to add more modules if needed, e.g., more sensors, without significant effort. The operating clock of the microcontroller is fast enough for these modules, thus they can be used at their specification limits.

The resulting measuring algorithm can be highly smart and automated. The microcontroller can implement a typical time reference and operate according to it. This time reference will be used in combination with possible external signals, like data from the computer or wireless signals through the RF transceiver. Also, external signals can exist of synchronizing nature, for combining multiple systems under study, or for combining this system with synchronized video data. So, the conditions for executing various operations create high automated algorithms for data measuring, and the time reference of these measurements can be in relation with available data obtained by outside systems, leading in total to good synchronized data.

The combination of high functionality and high portability makes the measuring procedure quite easy to set-up and handle. An issue on video analysis systems is the demand of highly specialized personnel in terms of correctly setting up the experiment (e.g., accurately placing marks on the human body, complicated calibration procedures). These systems have equally difficult software running on a computer. On the other hand, the system under study does not need advanced training for a person in order to operate it, thus making the conduction of experiments easier.

8.3.6 Reliability — Self Test

The existence of a microcontroller can also add functionality to the complete system related to its good operating condition and the reliability of the measurements. First, the system can execute algorithms that check the good condition of the sensors. Internal references can be selected on them and compare the resulting measurements with previously obtained reference measurements. This procedure may indicate a possible malfunctioning of a sensor module. Also, procedures may be applied that use the memory module and the wireless transmitter. These procedures will utilize these modules in a repeating way, with responses from these modules that will also have been previously recorded. The comparison of every new response with an already recorded one will provide a means of verifying at some level the good operating condition of these modules.

A more advanced verification of the good operating condition of the system can be made with the help of external equipment. The research team of Konstantakos et al., 2007 [15] has provided a way of measuring the exact energy consumption of specific modules that will execute specific or random functions, by just recording the current drawn by the complete system's power supply. This measurement demands external hardware, although it is possible to include the energy measuring hardware as a module, and thus have it operate all the time. For this measurement, all modules must be placed under specific operating conditions and the resulting current is expected to be similar with an already measured value. Possible different values may indicate a possible malfunction of a module. This whole self-test functionality can be automated, which significantly enhances the reliability of the system under development.

8.3.7 Cost

Motion analysis systems based on such sensors are generally a good choice in economic terms since they use low cost hardware [16]. Accelerometers, gyroscopes and magnetic sensors are available in the market at reasonable prices. Even the commercial motion analysis packages based on such sensors are in very competitive prices.

The proposed hardware also consists of rather inexpensive modules, widely available on the market. The construction of the complete system consists of simple PCB construction, which can be made in any elementary electronics laboratory. The development environment that runs on the computer can be free of charge for a developer. In total, the complete system in all possible steps of development is inexpensive, especially when compared to the already known kinematic analysis systems that exist on the market, which are based on conventional video, or on the infrared optoelectronic technology.

8.3.8 Limitations

Besides the advantages that may arise from the above characteristics, the proposed system comes also with some limitations. One of the limitations, as far as the measurement procedure is concerned, is the need of a global reference. The system is able to record movement but is unaware on any initial orientation. The problem can be solved by the use of the accelerometer and the direction of the gravity's g acceleration, but this is accurate only for static initial conditions. A magnetic sensor that can provide an absolute reference, regardless of the set up or the movement characteristics can resolve this matter in the future.

Another issue comes with the alignment of at least one sensor's axis to an anatomical axis of the body segment, which is necessary in order to provide comprehensive kinematic data of the segment's movement. This can be improved slightly by implementing smaller sized systems, but it relies mostly on the proper setting of the experiments.

It should be again pointed out that the sampling frequency that someone would suspect as a possible limitation is not an issue in such applications, because biological signals like those of the human movement are characterized by low frequencies in general. Thus the current sampling frequency is adequate for such applications.

As far as more technical matters are being addressed, there is an increased difficulty that occurs in case that more modules must be used, due to the increased data and the need of synchronization. In this situation, a higher level control unit must be developed, with the ability to organize and to supervise the different modules being used. Also, the ability to store data on a memory card is convenient but a more real-time and without wires approach would enhance the system's possibilities. This can be realized by adding wireless communication to the system and having the data sent directly to a personal computer. Some preliminary work has been made in this way, indicating that such approaches are possible in the future.

One major drawback of the sensors based systems for their use for the study of human movement is the absence of video image. Video information is essential when it comes to human motion. The ability to combine video information is important as we mainly comprehend movement patterns through vision. Vision is the most important sense. From simple people to specialists (biomechanists, orthopedics) video is used for confirming or/and understanding the rational of numbers. The ultimate test for movement simulation models that are used in robotics, prosthetics or virtual reality modules is to make the movement look "natural". The first screening of a musculoskeletal pathology is not the diagrams of a gait analysis but the visual observation of the patient. Clinicians take no decisions without seeing the patient. Thus standalone the sensor only built system is not a conclusive solution without video/image screening.

Up to this point the major aspects of the sensors based systems in general and the proposed system, in particular, are covered. The key points have been presented. The following step is the presentation of a case study with measurements taken by the implemented system.

8.4 Case Study

The proposed system was used in a case study from the orthopedics field. Orthopedics is a field of medicine where the movement analysis and especially gait analysis is extensively used as a decision making tool and it is considered a step forward to the "evidence based medicine" approach.

More specifically a woman (65 years old) with a partial caseation of her left ankle was asked to walk for several steps (>15). A gyroscope and an accelerometer were mounted on her shank along the longitudinal axis of the tibia. The second pair of sensors was mounted on the 3rd tarsal of her foot.

For the presented case study, video was used as a qualitative tool. Video information, even if it is for qualitative purposes must be synchronized with the rest of the data. The synchronization of video information with the sensors based system was made possible through flashing leds mounted on the system (Figure 8.4), which were visible from the video capture source and since the system can accept such signals, were also monitored by the microcontroller and stored in memory.

The gait cycle step markers were calculated for the accelerometer data while the ankle joint angle was derived by integration of the gyroscope data. Results of the left and right ankle joint angle are illustrated in Figure 8.5.

They clearly indicate the deterioration of the range of motion of the left ankle. The reduced mobility of the left ankle is also imprinted in the angular displacement-angular velocity plots of the shank and foot segment [17], as they appear in Figures 8.6 and 8.7.

Both segments of the affected side have reduced range of motion and reduced angular velocity. Those data are very useful for orthopedics as they quantitatively reflect the underlying pathology, though a simple data

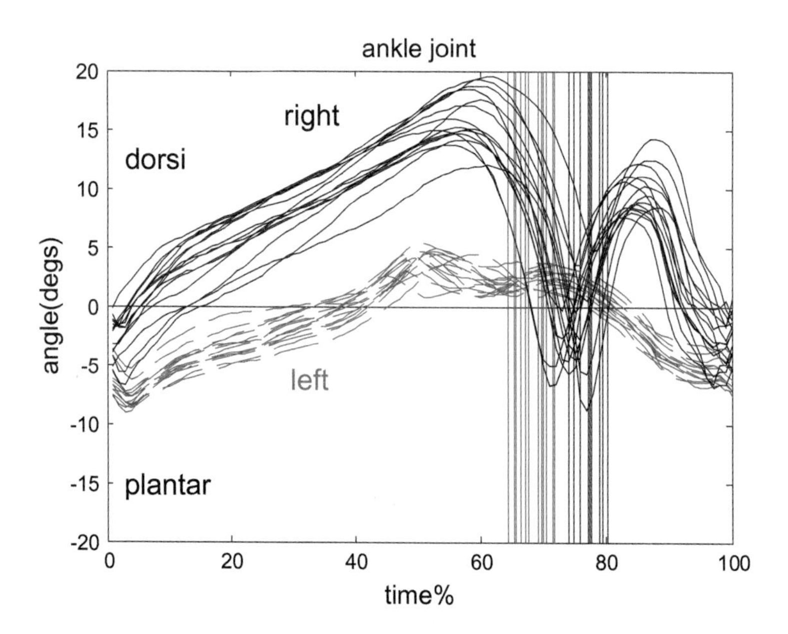

Fig. 8.5 Gait cycle of ankle joint angle.

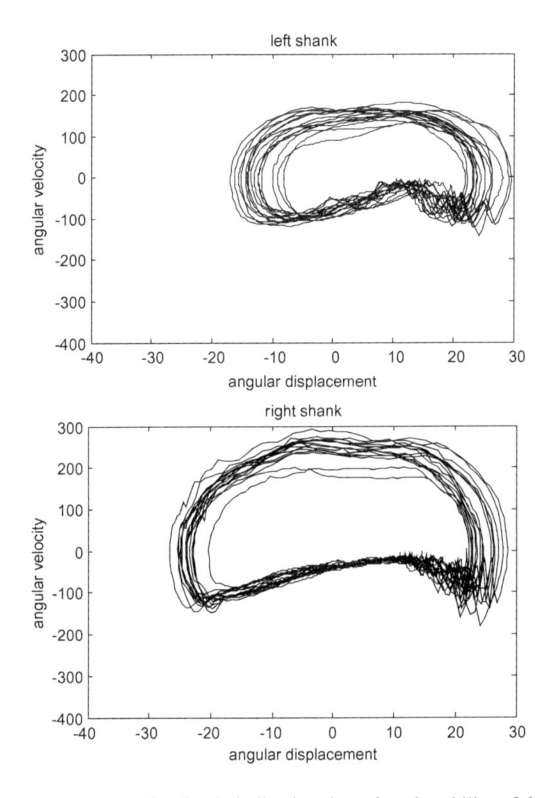

Fig. 8.6 Measurements on the shank, indicating the reduced mobility of the left ankle.

processing method for the user. Even more they serve the purpose of "evidence based medicine" the lack of which is a short come in many pathological cases on orthopedics [18]. At the present case, the doctor of the patient studied the curves before deciding to operate the patient. Both the patient and the doctor will be able to evaluate the result of the surgery not only subjectively but also objectively, based on a post-surgery measurement with the proposed system.

8.5 Future Research

If the sensors based systems for motion analysis continue to gain ground in the market then the demand to combine a large number of sensors for the implementation of full body or complex detailed analysis models will be prominent. Challenges will appear concerning the synchronization of all the sensors data [19] and the handling of a rather significant data rate, especially

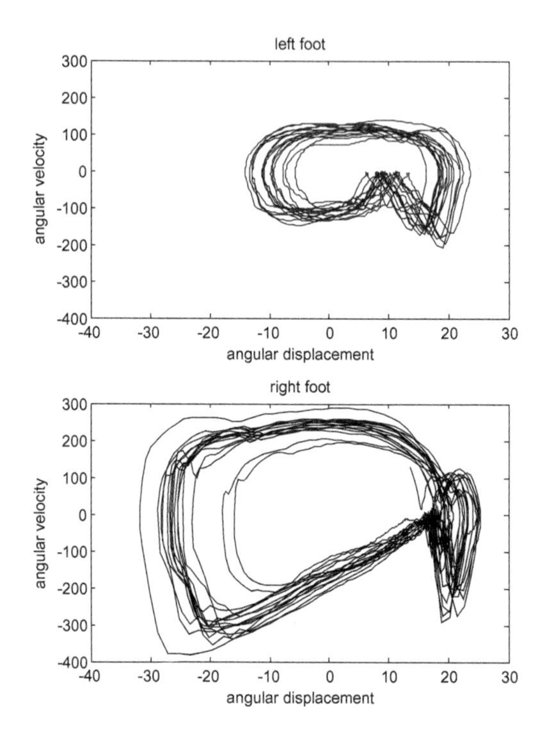

Fig. 8.7 Measurements on the foot, indicating the reduced mobility of the left ankle.

as the sampling frequency increases. Proper multiple sensors protocols will have to be implemented that will provide accurately synchronized data, easily gathered at a PC, and will also give the ability to connect/combine modules at a convenient way, thus making the experimental set-up easy to create and experiment with. Wired and wireless communication will become the means for synchronization. More complicated processing modules will be explored for implementing the main control unit of the complete system and for gathering all acquired data. Furthermore, multi-client communication protocols [20], or more generalized wireless sensors application platforms [21] will be developed to facilitate the programming needs.

In general the ability to add sensors in a protocol is important when it comes to study complex movements. This achievement will make possible practically every kind of movement analysis, by implementing primary a full body analysis solution with all the adequate detail and further more a custom need based solution, built on demand.

Since the use of the video (image) is fundamental for the comprehension of movement analysis, a system that will comprise both sensors and video can have many advantages. The combined system will have direct/precise measurements of position in space from video and velocity and acceleration measurements from the sensors. Video extracted measurements are comprised by the use of a low cost commercial camera and good algorithms for image analysis and data processing. On the other hand, sensors provide accurate data of the derivatives of motion. Thus the combined information will avoid the error of derivation and integration (numerical processing error) and with proper software, based on key events (the measured positions), all movement variables will be optimized (up-sampled, shift corrected), based on measured parameters. According to the aforementioned text that system could be a low cost yet powerful motion analysis solution.

References

[1] Y. I. Abdel-Aziz and H. M. Karara, "Direct linear transformation from comparator coordinates into object space coordinates," *Proceedings of the Symposium on Close Range Photogrammetry*, pp. 1–18, 1971.

[2] T. Nikodelis, D. Moscha, D. Metaxiotis, and I. Kollias, "Commercial video frame rates can produce reliable results for both normal and CP spastic gait's spatiotemporal, angular, and linear displacement variables", *Journal of Applied Biomechanics*, vol. 27, no. 3, pp. 266–271, 2011.

[3] A. Forner-Cordero, M. Mateu-Arce, I. Forner-Cordero, E. Alcántara, J. C. Moreno, and J. L. Pons, "Study of the motion artefacts of skin-mounted inertial sensors under different attachment conditions," *Physiological Measurement*, vol. 29, no. 4, pp. 21–31, 2008.

[4] K. J. O'Donovan, R. Kamnik, D. T. O'Keeffe, and G. M. Lyons, "An inertial and magnetic sensor based technique for joint angle measurement," *Journal of Biomechanics*, vol. 40, no. 12, pp. 2604–2611, 2007.

[5] K. Liu, T. Liu, K. Shibata, Y. Inoue, and R. Zheng, "Novel approach to ambulatory assessment of human segmental orientation on a wearable sensor system," *Journal of Biomechanics*, vol. 42, no. 16, pp. 2747–2752, 2009.

[6] G. Giakas and B. Baltzopoulos, "Optimal digital filtering requires a different cut-off frequency strategy for the determination of the higher derivatives," *Journal of Biomechanics*, vol. 30, no. 8, pp. 851–855, 1997.

[7] D. T. Fong and Y. Chan, "The use of wearable inertial motion sensors in human lower limb biomechanics studies: A systematic review," *Sensors*, vol. 10, no. 12, pp. 11556–11565, 2010.

[8] M. H. Schwartz and A. Rozumalski, "A new method for estimating joint parameters from motion data," *Journal of Biomechanics*, vol. 38, no. 1, pp. 107–116, 2005.

[9] D. A. Winter, *Biomechanics and Motor Control of Human Movement*, New York: Wiley & Sons, 2004.

[10] S. C. Stiros, "Errors in velocities and displacements deduced from accelerographs: An approach based on the theory of error propagation," *Soil Dynamics and Earthquake Engineering*, vol. 28, no. 5, pp. 415–420, 2008.

[11] C. Angeloni, P. O. Riley, and D. E. Krebs, "Frequency content of whole body gait kinematic data," *IEEE Transactions on Rehabilitation Engineering*, vol. 2, no. 1, pp. 40–46, 1994.

[12] V. Konstantakos, T. Nikodelis, I. Kollias, and T. Laopoulos, "Automated measurement system based on digital inertial sensors for the study of human body movement," *6th IEEE International Conference on Intelligent Data Acquisition and Advanced Computing Systems: Technology and Applications*, pp. 76–80, 2011.

[13] B. Coley, B. Najafi, A. Paraschiv-Ionescu, and K. Aminian, "Stair climbing detection during daily physical activity using a miniature gyroscope," *Gait Posture*, vol. 22, no. 4, pp. 287–294, 2005.

[14] V. Konstantakos, A. Chatzigeorgiou, S. Nikolaidis, and T. Laopoulos, "Energy consumption estimation in embedded systems," *IEEE Transactions on Instrumentation and Measurement*, vol. 57, no. 4, pp. 797–804, 2008.

[15] V. Konstantakos and Th. Laopoulos, "Self-evaluation configuration for remote data logging systems," *IEEE International Workshop on Intelligent Data Acquisition and Advanced Computing Systems: Technology and Applications (IDAACS)*, pp. 18–23, 2007.

[16] R. E. Mayagoitia, A. V. Nene, and P. H. Veltink, "Accelerometer and rate gyroscope measurement of kinematics: An inexpensive alternative to optical motion analysis systems," *Journal of Biomechanics*, vol. 35, no. 4, pp. 537–542, 2002.

[17] N. Stergiou, *Innovative Analyses of Human Movement*, Human kinetics, 2004.

[18] R. E. Cook, I. Schneider, M. E. Hazlewood, S. J. Hillman, and J. E. Robb, "Gait analysis alters decision-making in cerebral palsy," *Journal of Pediatric Orthopaedics*, vol. 23, no. 3, pp. 292–295, 2003.

[19] Y. C. Wu, Q. Chaudhari, and E. Serpedin, "Clock synchronization of wireless sensor networks," *IEEE Signal Processing Magazine*, vol. 28, no. 1, pp. 124–138, 2011.

[20] A. Chehri and H. Mouftah, "A practical evaluation of zigbee sensor networks for temperature measurement," Lecture Notes of the Institute for Computer Sciences, Social Informatics and Telecommunications Engineering, vol. 49, Part 9, pp. 495–506, 2010.

[21] S. Zhang, Z. Dong, Y. Cui, and Y. Dong, "A middleware platform for WSAN based application systems," *5th Intern. Conference on Computer Sciences and Convergence Information Technology (ICCIT)*, pp. 981–986, 2010.

9

Monitoring Assisted Livings through Wireless Body Sensor Networks*

Domenico Luca Carnì*, Giancarlo Fortino†, Raffaele Gravina‡, Domenico Grimaldi§, Antonio Guerrieri¶, and Francesco Lamonaca‖

Department of Electronics, Informatics and Systems (DEIS), University of Calabria, Via P. Bucci, 87036 Rende (CS), Italy;
**dlcarni@deis.unical.it; †fortino@unical.it; ‡rgravina@deis.unical.it; §grimaldi@deis.unical.it; ¶aguerrieri@deis.unical.it; ‖flamonaca@deis.unical.it*

Abstract

Wireless Body Sensor Networks (WBSN) are demonstrating to have the potential to effectively support a great variety of high-impact application domains, from e-Health to e-Factory. This chapter first overviews reference network architectures, effective programming frameworks and novel applications in important application domains for WBSNs. Then, it introduces SPINE (Signal Processing In-Node Environment), an effective and efficient framework for programming WBSN systems, and analyzes SPINE-based technology and methods for capturing and processing signals coming from the human body. Finally, a SPINE-based WBSN system in the e-Health application domain

*This work has been partially supported by CONET, the Cooperating Objects Network of Excellence, funded by the European Commission under FP7 with contract number FP7-2007-2-224053.

Advanced Distributed Measuring Systems — Exhibits of Application, 211–242.

concerning anytime and anywhere remote and synchronized monitoring of heart rate and activity of assisted livings is described and analyzed.

Keywords: WBSN, programming frameworks, e-Health.

9.1 Wireless Body Sensor Networks

Wireless Sensor Networks (WSNs) are currently emerging as one of the most disruptive technologies enabling and supporting next generation ubiquitous and pervasive computing scenarios [28]. In particular, Wireless Body Sensor Networks (WBSNs) are conveying notable attention as their real-world applications aim at improving the quality of human beings life by enabling continuous and real-time non-invasive assistance at low cost and effectively supporting emergency response scenarios.

A wireless Body Sensor Network (WBSN) is a collection of wearable (and programmable) sensor nodes communicating with a local personal device (see Figure 9.1). The sensor nodes have computation, storage, and wireless transmission capabilities, a limited energy source (i.e., battery), and different sensing capabilities depending on the physical transducer(s) they are equipped with. Common physiological sensing dimensions include body motion, skin temperature, heart rate, muscular tension, breathing rate and volume, skin conductivity, and brain activity. The local personal device is typically a

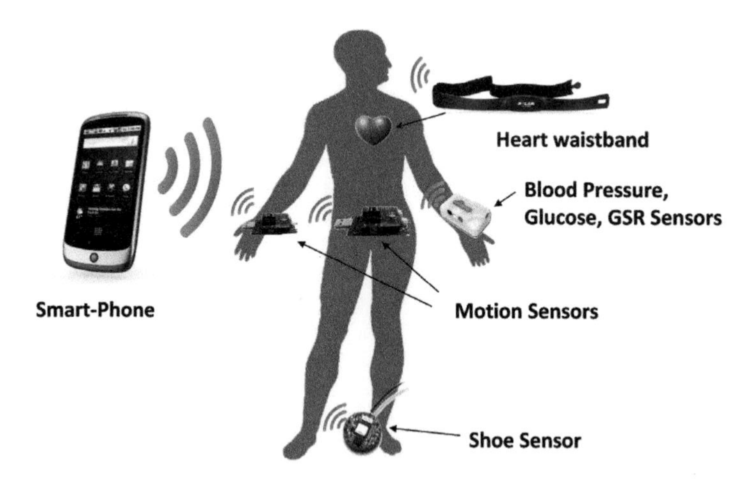

Fig. 9.1 WBSN Architecture.

smart-phone or a PC, and allows for real-time monitoring as well as long-term remote storage and off-line analysis.

However, designing and programming applications based on WBSNs are still complex tasks. That is mainly due to the challenge of implementing signal processing intensive algorithms for data interpretation on wireless nodes that are very resource limited and have to meet hard requirements in terms of wearability and battery duration as well as computational and storage resources. This is challenging because WBSNs applications usually require high sensor data sampling rates that endanger real-time data processing and transmission capabilities as computational power and available bandwidth are generally scarce. This is especially critical in signal processing systems, which usually have large amounts of data to process and transmit.

The remainder of the chapter is organized as follows. Section 9.1.1 introduces the main application domains in which WBSNs can be effectively involved. Section 9.1.2 briefly describes some examples of popular research and commercial WBSN systems. Section 9.2 overviews the most relevant related work on domain-specific frameworks for programming WBSN systems and describes in detail the architecture and main functionalities of the SPINE Framework. Section 9.3 describes the flexibility of the SPINE Framework showing examples of raw sensor data recording using SPINE, as well as analyzing some of its interesting in-node signal processing functionalities. Section 9.4 describes and analyzes a WBSN case study in the e-Health domain that allows for continuous and real-time monitoring of activity and heart rate of assisted livings. Finally, Section 9.5 includes some concluding remarks and comments on future research directions that can derive from the topics here presented.

9.1.1 Applications Domains

WBSNs can enhance many human-centered application domains. We can categorize them into the followings:

- e-Health;
- e-Emergency;
- e-Entertainment;
- e-Sport;
- e-Factory;
- e-Sociality.

E-Health applications span from early detection or prevention of diseases (heart attacks, Parkinson, diabetes, etc.), elderly assistance at home, e-fitness, post-trauma rehabilitation after surgeries, motion and gestures detection, cognitive and emotional recognition, medical assistance in disaster events [8].

E-Emergency applications include WBSN systems to support fire fighters, response teams in large scale disasters due to earthquakes, landslides, terrorist attacks, etc [14].

E-Entertainment domain refers to human-computer interaction systems typically based of WBSNs for real-time motion and gesture recognition [20].

E-Sport applications are related to the e-Health domain, although they have a non-medical focus. Specifically, this domain includes personal e-fitness applications for amateur and professional athletes, as well as enterprise systems for fitness clubs and sport teams offering advanced performance monitoring services for their athletes [7].

E-Factory is an emerging and very promising domain in which WBSNs have a central relevance; applications in such domain aim at monitoring employees' activities, such as in production chains, to both help ensure safety and to guide proper assembly of the product [12].

Finally, e-Sociality applications may use WBSN technologies to recognize user emotions and cognitive states to enable new forms of social interactions with friends and colleagues. An interesting example is given by a system that involves the interaction between two people's WBSNs to detect handshakes and, subsequently monitor their social and emotional interactions [3].

9.1.2 Research Projects and Commercial Products

Most of the literature on WBSN systems focuses on the functional and system perspective, and a few details are typically given on the software design and architecture; sometimes the software side of the system is completely left uncovered. In the absence of a systematic description of such aspects, we have to assume that the software engineering has not been given enough importance while developing the system. Hence, we can definitely state that currently most of the WBSN implementations are based on application-specific code.

In the following, two recent WBSN research projects and two WBSN-inspired commercial products will be briefly introduced. A summary of some

Table 9.1. Summary of some well-known WBSN systems.

Project Title	Application Domain	Sensors Involved	Hardware Description	Node Platform	Communication Protocol	OS / Programing Language
Real-time Arousal Monitor	Emotion recognition	ECG, Respiration, Temp., GSR	Chest-belt, skin electrodes, wearable monitor station, USB dongle	custom	Sensors connected through wires	n.a. / C-like
LifeGuard	Medical monitoring in space and extreme environments	ECG, Blood Pressure, Respiration, Temp., Accelerometer, SpO_2	Custom microcontroller device, commercial bio-sensors	*XPod* signal conditioning unit	*Bluetooth*	n.a.
Fitbit®	Physical activity, sleep quality	Accelerometer	Waist/wrist-worn device, PC USB dongle	*Fitbit®* node	RF proprietary	n.a.
VitalSense®	In-and on-body Temperature, Physical Activity, Heart monitoring	Temp., ECG, Respiration, Accelerometer	Custom wearable monitor station, wireless sensors, skin electrodes, ingestible capsule	*VitalSense®* monitor	RF proprietary	Windows Mobile
LiveNet	Parkinson symptom, epilepsy seizure detection	ECG, Blood Pressure, Respiration, Temp., EMG, GSR, SpO_2	PDA, microcontroller board	custom physiological sensing board	wires, 2.4GHz radio, GPRS	Linux (on PDA)
AMON	Cardiac-respiratory diseases	ECG, Blood Pressure, Tem., Accel., SpO_2	Wrist-worn device	custom wrist-worn device	Sensors connected through wires - GSM/UMTS	C-like / JAVA (on the server station)
MyHeart	Prevention and detection of cardio vascular diseases	ECG, Respiration, Accelerometer	PDA, Textile sensors, chest-belt	Proprietary monitoring station	conductive yarns, Bluetooth, GSM	Windows Mobile (on the PDA)
Human++	General health monitoring	ECG, EMG, EEG	Low-power BSN nodes	ASIC	2.4GHz radio / UWB modulation	n.a.
HealthGear	Sleep apnea detection	Heart Rate, SpO_2	Custom sensing board, commercial sensors, cellphone	custom wearable station	Bluetooth	Windows Mobile (on the mobile phone)
TeleMuse®	Medical care and research	ECG, EMG, GSR	*Zigbee* wireless motes	proprietary	IEEE 802.15.4 / *Zigbee*	C-like
Polar® Heart Rate Monitor	Fitness and exercise	Heart Rate, altimeter	Wireless chest-belt, watch monitor	proprietary watch monitor	*Polar OwnCode®* (5 kHz) – coded transmission	n.a.

literature WBSN systems is reported in Table 9.1. A comprehensive overview of several WBSN applications can be found in [1, 10].

9.1.2.1 Real-time Arousal Monitor

An interesting research work regarding the design of a real-time arousal monitor has been presented in [9]. *Arousal* is a physiological and psychological state of being awake or reactive to stimuli. The goal of the study aims to monitor the level of arousal on a continuous scale, and continuously in time. It is different from other works in that it focuses on light-weight algorithms that yield a continuous estimation of the non-discretized arousal level, having the potential of being integrated as a part of the WBSN.

The system detects four signals that are known to be directly influenced by a subject's state of arousal through the activation of the *Autonomic Nervous System* (ANS) — being ECG, respiration, skin conductance, and skin temperature. ECG is recorded through a proprietary chest belt for bio-potential read-out. Also integrated in the belt is a piezoelectric film sensor used to measure respiration. Skin conductance is measured at the base of two fingers, by measuring the electrical current that flows as a result of applying a constant voltage. Skin temperature is measured at the wrist, by using a commercially available digital infrared thermometer module.

Data is received wireless by the base-station that is connected to a PC via USB. An interface to Matlab has been realized to enable a platform for quick development and verification of real-time physiological signal processing algorithms. Interpolation of missing data is also present.

The main weakness of this proposal is the lack of a good reference which makes difficult to quantify the quality of the analysis. Furthermore, all experiments have been performed in a controlled environment. Further experiments are needed before conclusions can be drawn about the extension of the results to non-controlled environments.

9.1.2.2 LifeGuard

Another research work that is worth mentioning is *LifeGuard* [15]. The goal of this effort was to design a small, light weight, wearable, ergonomic device for a NASA research that not only records and streams a comprehensive set of diagnostic-quality physiologic parameters, but can also record body position and orientation, acceleration in three axes, and can be used to mark events. This feature set, combined with wearability, alarm indicators, fault detections, and the ability to stream data to hand-held Bluetooth-enabled devices, forms a compact and reliable system.

The LifeGuard system consists of the "Crew Physiologic Observation Device" (CPOD) and a portable base-station computer. The CPOD device, the core component of the system, is a custom-made, small, lightweight, easy-to-use device that is worn on the body along with the physiologic sensors described below. It is capable of logging physiologic data as well as transmitting data to a portable base-station computer for display purposes and further processing. Most physiologic parameters (ambient and skin temperature, ECG, respiration rate, pulse oximetry, blood pressure) supported by LifeGuard are measured with sensors that are external to the CPOD wearable device, and that can be connected to it via wired plugs. The only sensors that are integrated into the CPOD are the accelerometers.

The user cannot change the basic operational mode of the CPOD. The authors motivate this design choice as an additional security so that the system meets the high reliability standards of medical monitoring devices. However, the 6 hours streaming (or 20 hours logging) battery lifetime make LifeGuard hardly usable in non-critical environments. Additionally, as the system cannot be re-configured easily, re-programming is necessary every time some of the tunable system properties (the sensors to enable, or the sensor sampling rate to name a few) need to be modified.

9.1.2.3 FitBit

Fitbit [21] is a promising commercial product thanks to its relatively low cost, the various supported features, and its graceful cosmetic design.

Fitbit consists of a wearable small sensor device, a base-station connected to a PC, and a web-based application used to record, visualize and analyze collected data. Through an embedded accelerometer, Fitbit allows the estimation of the number of steps taken during walking or running, the distance traveled, the calories burned during physical activities such as walking, jogging, and other daily life activities. It provides statistics about daily levels of activity (sedentary, lightly active, fairy active, very active). All these information are available in real-time on a tiny OLED display placed on one side of the wearable device. Fitbit also gives information about the quality of sleep (e.g., how long did it take to fall asleep, how long was the sleep, how many times the user has awakened).

To work properly, it must be worn on the waist or the chest, or, while sleeping, with the provided wristband. Recorded data are temporarily stored locally up to 7 days, and transmitted via wireless to the base-station automatically as soon as the wearable device comes in the range of 15 feet. The declared battery lifetime is 5–10 days of continuous use. Because each device has an unique ID, a single base-station can be used to gather information from multiple Fitbit devices in the nearby. An Internet connection is needed to send the recorded data to a dedicated server. An intuitive and user-friendly web-based application allows the user tracking historical statistics, and visualizing graphically his/her personal progress from time to time. Although the system does not provide any way to dump data to the PC, the website will have an extensive XML and JSON API to access most of stored data.

9.1.2.4 VitalSense

Another commercially available physiologic monitoring system is *VitalSense* [26]. The VitalSense system includes different types of wireless sensors, a small monitor equipped with control buttons and a graphical display, and a software running on the PC.

VitalSense is designed to monitor temperature, in active or inactive subjects and in indoor and outdoor environments. Each VitalSense monitor can receive transmissions from up to 10 miniature, wireless, temperature sensors. Core temperature is sensed by small ingestible capsules. Dermal temperatures are recorded from hypoallergenic adhesive dermal patches. Both sensor types are disposable, but designed for multi-day use under demanding physical and environmental conditions. In addition, the VitalSense telemetric physiological monitoring system can be interfaced to the so called *VitalSense-XHR* sensor: a compact device that transmits over the air Heart Rate and Respiration Rate (which is derived from the ECG). This sensor enables researchers to monitor heart rate and respiratory rate on moving subjects. The XHR is a chest-worn wireless water-resistant physiological monitor that incorporates an ECG-signal processor. The rechargeable battery in the XHR provides four days of battery life on a full charge.

Furthermore, VitalSense can also be interconnected with the *Actical* device. Caloric expenditure, and counting of the steps taken during walking are calculated from movement recorded by the Actical device worn on the waist.

VitalSense is supported by a software utility that communicates with the portable monitor station via an RS-232 cable. The software enables to setup the monitor and initialize the sensors. It also allows real-time temperature data monitoring as well as retrieval — and off-line visualization — of recorded data from the monitor.

9.2 The SPINE Framework

SPINE [4, 5, 23] is an application level domain-specific open-source framework for fast prototyping of applications based on WBSNs. SPINE provides support to distributed signal-processing intensive WBSNs applications by a wide set of pre-defined physiological sensors, signal-processing utilities, and flexible data transmission. Furthermore, it has a powerful and well designed modular structure that allows for easy integration of new custom-designed sensor drivers and processing functions, as well as flexible tailoring and customization of what is already supported, to fit specific developer needs. One of the fundamental ideas behind SPINE is the software components reuse to allow different end-user applications to configure the sensor nodes at runtime based on the application-specific requirements, so that the same embedded code can be used for several applications without re-programming off-line the sensor nodes before switching from an application to another.

9.2.1 Related Work

Programming WBSN systems is a complex task. That is mainly due to the challenge of implementing signal processing intensive algorithms for data interpretation on wireless nodes that are very resource limited and have to meet hard requirements in terms of wearability and battery duration as well as computational and storage resources. As a consequence, to support application developers with higher abstractions than what is usually provided natively by the specific platform APIs, a number of domain-specific software frameworks for WBSNs have been proposed so far. Domain-specific frameworks are novel software systems following an approach in the middle between application-specific code and general-purpose middleware. They specifically address and standardize the core challenges of WSN design within a particular application domain. While maintaining high efficiency, such frameworks allow for a more effective development of customized applications with little or no additional hardware configuration and with the provision of high-level programming abstractions tailored for the reference application domain.

Specifically, besides SPINE, the most relevant frameworks for WBSNs are: CodeBlue, Titan, RehabSPOT, and the most recent SPINE2.

CodeBlue [14] was designed to address a wide range of medical scenarios, such as monitoring patients in hospitals or victims of a disaster scene, where both patients/victims and doctors/rescuers may move and not necessarily be in direct radio range all the time. CodeBlue supports TinyOS-based sensor platforms.

Titan [13] is a general-purpose middleware that supports implementation and execution of context recognition algorithms in dynamic WSN environments. Titan represents data processing by a data flow from sensors to recognition result. It has been designed to run on TinyOS.

RehabSpot [29] is a customizable wireless networked body sensor platform for physical rehabilitation. RehabSPOT is built on top of SunSPOT technology [24] from Sun Microsystems. RehabSPOT-based WBSNs run a uniform program on all wearable nodes although they may perform different functions during runtime. The system software is based on client-server architecture. The server program is installed and running on the PC while the client program is installed in the remote nodes.

SPINE 2 [18] is a complete platform-independent re-engineering of the SPINE framework which implements a full task-oriented approach. It supports TinyOS [25] and Z-Stack based sensor nodes.

9.2.2 High-level Software Architecture

The SPINE framework consists of two main components:

(1) *SPINE Node*, and
(2) *SPINE Coordinator*.

The former is implemented in the sensor platform-specific embedded programming language and runs on the sensor nodes; the latter is implemented in Java and runs on the coordinator station. The functional architecture of SPINE is shown in Figure 9.2.

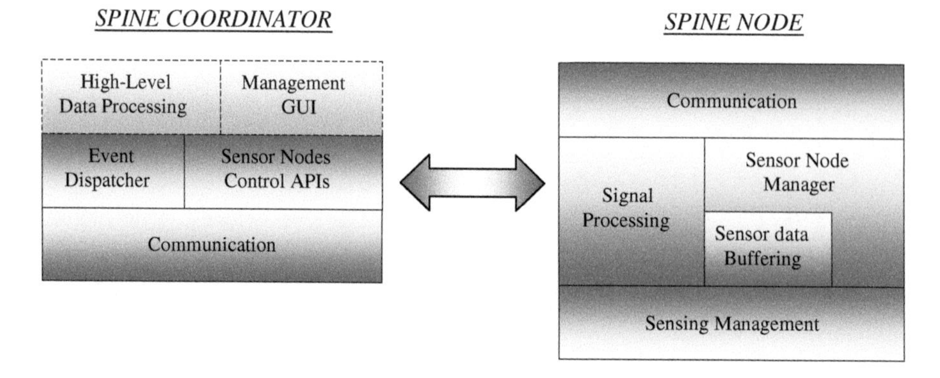

Fig. 9.2 The SPINE high-level functional architecture.

The *SPINE Node* is organized in five interacting functional components:

- The "*Sensor Node Manager*" is responsible for the general interactions among the Sensing Management, Signal Processing, and Communication components, and dispatches requests coming from the remote coordinator to the appropriate block.
- The "*Communication*" block handles reception and transmission of messages over-the-air, and managing radio duty cycling.
- The "*Sensing Management*" block acts as a general interface to the physical sensors of the platform, setting up timers when periodical sensing is requested by the remote coordinator, or simply performing one-shot reading to the requested sensors. It interacts with the "Buffering" block to store the sensor readings. It also contains a sensor registry where compiled sensor drivers self-register at boot-time, to allow other components to retrieve the available sensor list.
- The "*Buffering*" block consists of a set of circular buffers dedicated to store the sensor sample data. It provides two mechanisms to access the sensor data: (i) upon requests using getter functions, and (ii) using listeners that are notified when new data from sensor(s) of interest is available.
- The "*Signal Processing*" block involves a flexible, customizable and expansible set of signal processing functionalities such as math aggregators, filters, and threshold-based alarms that can be applied to any sensor data streams. This block retrieves the data to be processed from the sensor buffers, and report the results back over-the-air to the coordinator.

The *SPINE Coordinator* is organized in five components:

- The "*Communication*" block has similar functionalities of the corresponding block in the sensor node, and it has been designed to load the proper radio module adapter (e.g., for TinyOS motes or SunSPOT devices) dynamically. This component has the important function of abstracting the logical interactions between the coordinator and the WBSN from the low level transmissions that depend on the actual platform technology being used.

- The "*Sensor Nodes Control APIs*" is an interface exposed to the end-user application developers to manage the underlying WBSN (e.g., to activate sensor sampling and on-node signal processing to certain nodes).
- The "*Event Dispatcher*" forwards various events (e.g., new nodes discovered, alarm or user data messages) to the set of registered listeners.
- The "*High-Level Data Processing*", which is described in more details in paragraph 9.3.3, enables signal processing and pattern recognition on the coordinator node. Using the SPINE distributed computing architecture, this important module supports the design and implementation of new applications by providing highly generalized interfaces for data pre-processing, feature extraction and selection, signal processing, and pattern classification. It is designed to simplify the integration of SPINE in signal processing or data mining environments, providing functionality such as automatic network configuration, aggregate data collection, and graphical configuration. It also includes a bridge to the popular WEKA [11] Data Mining toolkit to allow the use of its feature selection and pattern classification algorithms from within SPINE.
- The "*SPINE Management GUI*" is a graphical add-on tool that allow configuration of remote sensor nodes using a user-friendly interface (rather than by manual coding). It contains a simple textual logging function for events generated by remote nodes and received by the underlying SPINE coordinator (e.g., discovery advertisement packets, data messages).

9.3 Biophysical Data Acquisition and Processing Using SPINE

9.3.1 Supported Biophysical Sensors

The SPINE Framework supports most of the sensor node platforms commonly used for WBSN applications. For instance, MicaZ, TelosB/Tmote Sky, and Shimmer (all versions) motes are fully supported. For each of these platforms, most of the built-in and optional physical sensors have been integrated and available for remote data collection. Specifically, the following biophysical and environmental information can be extracted with the open-source release of SPINE:

- body motion and postures (through accelerometers and gyroscopes);
- skin and environmental temperature;
- cardiac activity (through electrocardiography — or ECG — sensor);
- blood oxygen saturation, heart rate, and blood pressure (through photo-plethysmography — or PPG — sensor);
- respiration rate (through electro-impedance-plethysmography — or EIP — sensor);
- skin resistivity (through galvanic skin response — or GSR — sensor);
- humidity;
- light.

Obviously, each of the listed sensors is not typically available on every supported platform.

Furthermore, the node-side software infrastructure has been carefully designed to realize an abstraction layer of the sensing functionalities so that it is easy for developers to integrate further sensor drivers and it is even more simple for coordinator-side developers to acquire any kind of raw sensor data.

In the following, some practical examples of biophysical signals that can be collected at the coordinator using the very intuitive Java-based SPINE API [5] are provided.

Figure 9.3 shows frontal, vertical, and longitudinal acceleration of the waist during a short walk on a straight and plain path. A simple data collection application uses the SPINE coordinator API to gather raw readings from a Tmote Sky node equipped with a custom motion-board to which a three-axis accelerometer is attached.

The same data collection application has been used to acquire the ECG signal of a healthy subject during resting (see Figure 9.4). Adapting the collection application to fit the ECG signal only required to change the sensor type and sampling frequency of the sensing request command provided by the SPINE framework API.

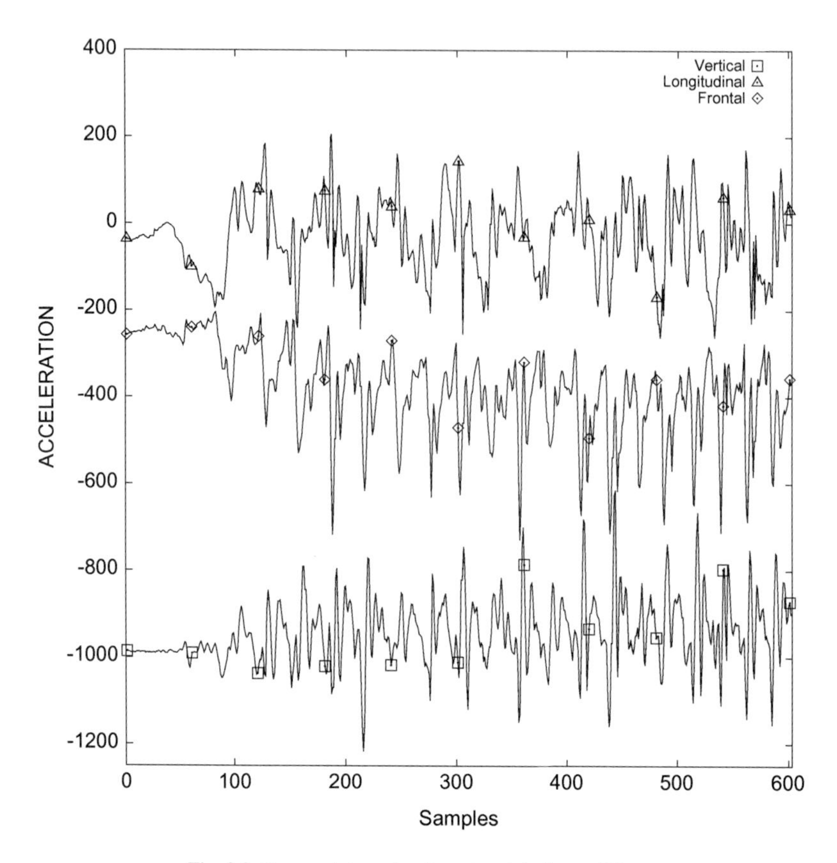

Fig. 9.3 Raw waist acceleration signal during walking.

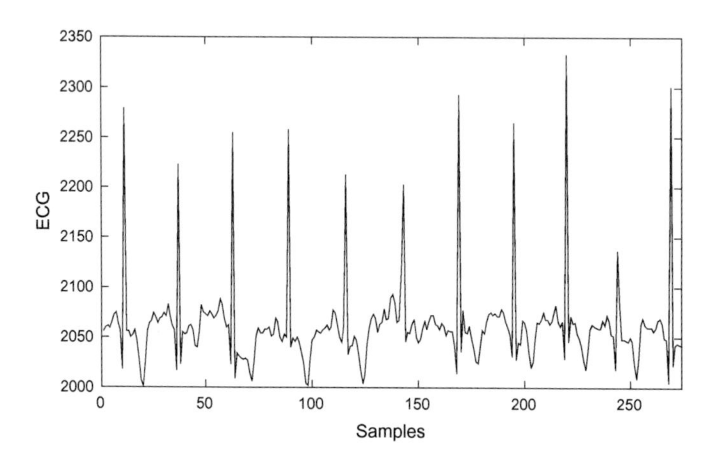

Fig. 9.4 ECG signal of a healthy subject during resting.

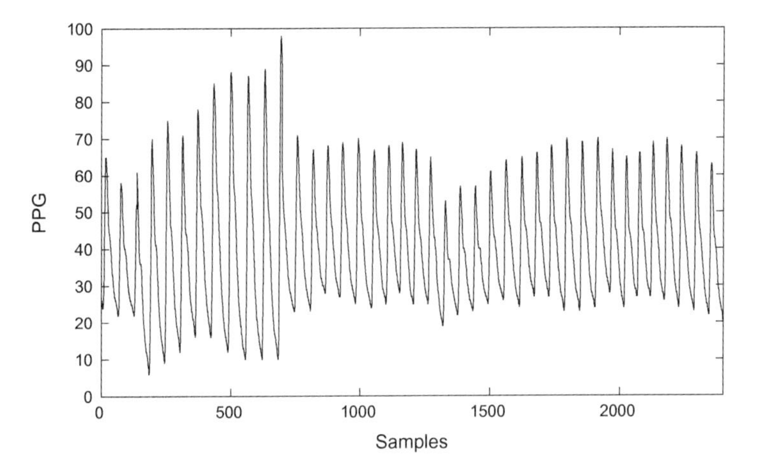

Fig. 9.5 PPG signal of a healthy subject while seated.

For this experiment, we used a Shimmer 2R that has been extended with the ECG sensor-board which is commercially available at the Shimmer website [22].

Finally, the data collection application has been configured to acquire the PPG signal of another subject while seated over a period of 40 seconds (see Figure 9.5).

A PPG sensor is an optical device that can measure the change in blood relative volume [30] within body tissue with a non-invasive technique. It is based on the absorption spectroscopy, which is a technique for the chemical characterization. This method can be used to detect luminescent and non-luminescent materials and provide specific chemical information [27].

For this experiment, we used a custom-made PPG sensor attached to a Shimmer 2R node. The PPG sensor is composed by two parts:

1. the light source constituted by infrared (IR) diode;
2. the sensor realized by photo-detector.

In particular, for this specific application, it is important to observe that the living tissues are transparent to red and infrared light while non-hemolized blood is relatively opaque in this spectral range. As the blood absorbs or reflects part of the light, the variation of the blood volume caused by heart beats modulates the amount of transmitted or reflected light. This information can be used to infer several parameters regarding the dynamic activity of the vascular system.

The operation principle of the sensor applied to a tissue area can be simplified in the following steps:

- the diode emits a light beam into the skin towards the photo-detector;
- the hemoglobin into the blood absorbs part of the infrared light beam;
- the photo-detector converts the reflected or transmitted light spectrum in electrical signal.

To obtain accurate measurements, it is necessary to apply the sensor in arterio-venous rich areas such as the fingers, toes, earlobes, or some regions of the face. Although useful for the peripheral blood flow, the existing PPG devices are not suitable for measurements from the main body arteries due to the noise caused by muscle movements [30]. The light source characteristics influence the result of the measurements. Using an emitter with principal spectral output near 940nm allows to obtain measurements that are not influenced by the variation of oxygen saturation.

9.3.2 Enabling in-node Signal Processing

In Section 9.3.1 we emphasized the level of heterogeneity of the SPINE Framework in terms of supported biophysical sensors. As aforementioned, SPINE allows for retrieving raw sensory data from a remote coordinator through an high-level and intuitive Java-based API.

However, one of the most important characteristics of SPINE is the support for different types of signal processing functionalities that are executed directly on the sensor nodes and that can be dynamically enabled and disabled upon request of the end-user application running of the coordinator device.

Most of the natively supported signal processing is general purpose in the sense that can be applied to any of the available sensor signals, although a few functionalities are meaningful only if applied to specific sensory data.

SPINE supports many mathematical aggregators (also called “features”) that can be computed periodically on arbitrary sensor data windows, different types of threshold-based alarms that are triggered only if the signal to which they are been applied exceeds the defined threshold range, and a few simple digital filters. For a comprehensive description of the in-node signal processing functionalities we refer to [5].

In the following, we show significant features that can be computed on the sensory data depicted in Section 9.3.1. The SPINE nodes were configured to transmit both the raw sensor signals and the computed features so that we could show both on the same plot.

Figure 9.6 shows the output of a 5-point smoothing average Finite Input Response (FIR) filter applied to the accelerometer signals directly on the sensor node. By comparing the plot with the corresponding raw sensor readings shown in Figure 9.3, it is clear the "low-pass" effect of this filter.

Among the other in-node features, SPINE provides a peak detection which can be very useful for many applications. For instance, Figure 9.7 shows how

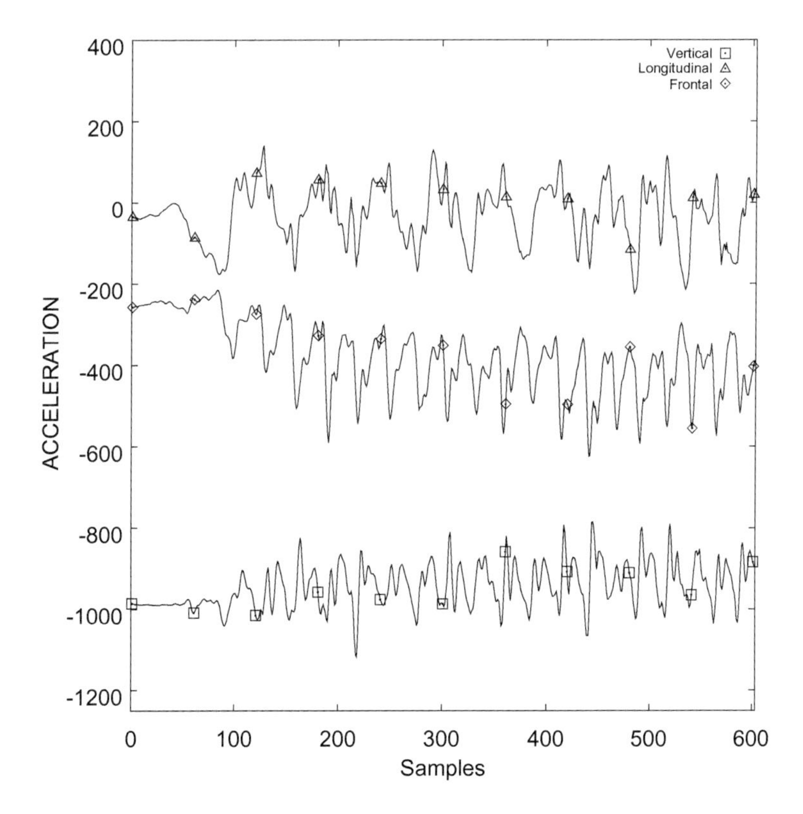

Fig. 9.6 Smoothing average FIR filter applied to accelerometer signals.

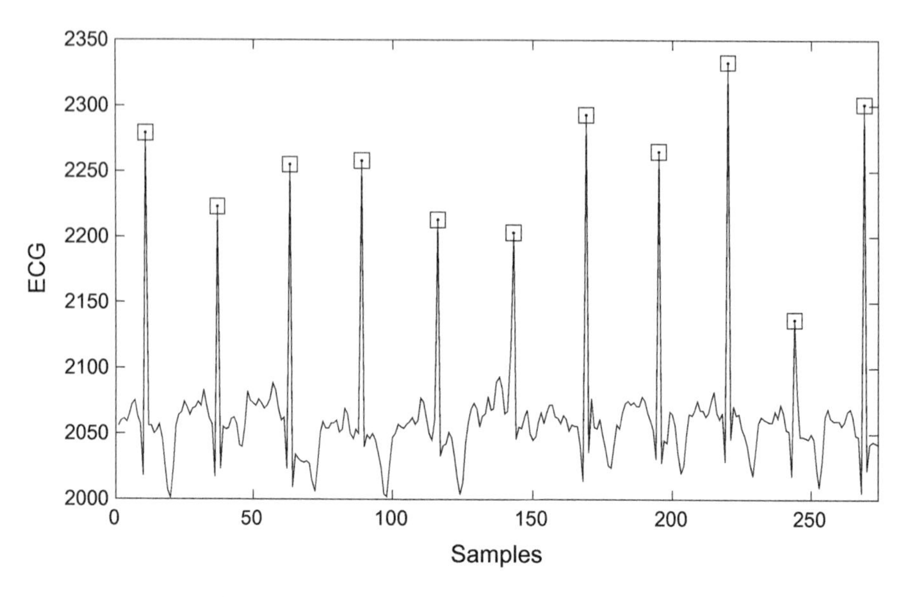

Fig. 9.7 Peak detection feature applied to ECG signal.

this can be useful to detect the peak of the QRS complex in the ECG signal which in turn allows to measure the heart rate (HR).

If a certain application only requires the HR, SPINE allows for transmitting a timestamped trigger message only when a heart beat is detected, which is much more convenient in terms of power consumption and bandwidth usage than transmitting periodically the full ECG back to the coordinator.

Figure 9.8 shows the local maximum and minimum points extracted from the PPG with the max and min feature extractors provided by SPINE.

The automatic detection of maximum and minimum points allows to determine and monitor the follow parameters:

- heart rate;
- sistolic time, which is defined as the ascending time of the signal from its minimum to its maximum (see Figure 9.9);
- diastolic time, which is defined as the descendant time of the signal from its maximum to its minimum (see Figure 9.10).

According to Teng and Zhang [19], these parameters could be used to monitor the Blood Pressure (BP) in an automatic and non-invasive modality.

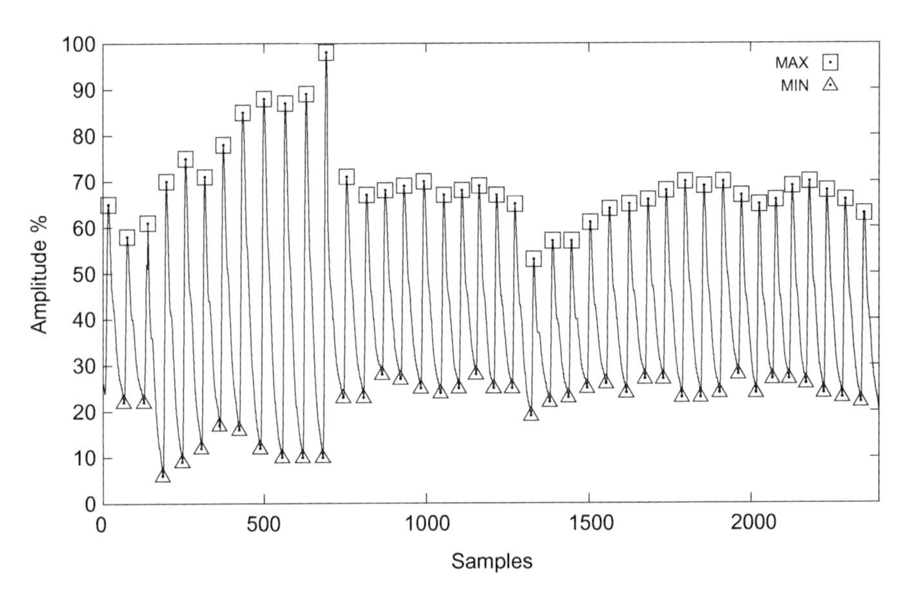

Fig. 9.8 PPG signal and detection of maximum and minimum points.

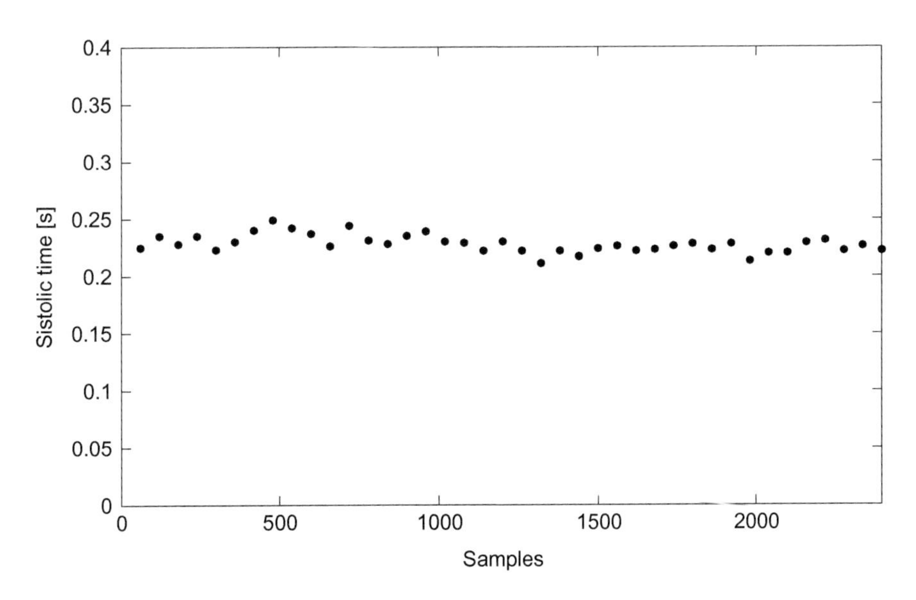

Fig. 9.9 Temporal characteristic of the PPG signal: *systolic time.*

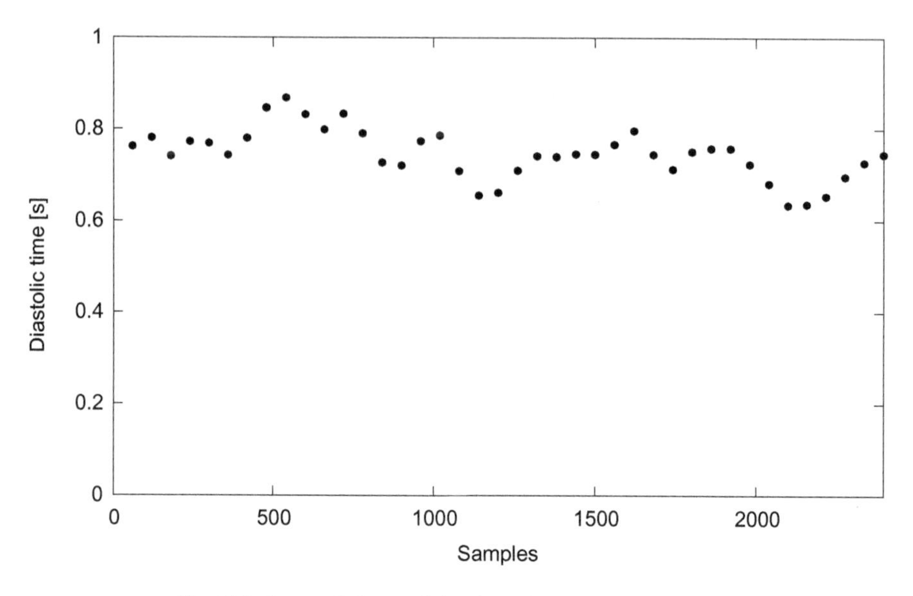

Fig. 9.10 Temporal characteristic of the PPG signal: *diastolic time*.

9.3.3 From Sensor Data Acquisition to Classification Results

Recently, SPINE has been extended with an optional plug-in which represents a powerful extension to the core framework as it provides more complex signal processing and decision support functionalities (e.g., classification) that are intended to be performed at the coordinator. It is designed to simplify the integration of SPINE in Signal Processing or Data Mining environments providing more application-oriented functions such as automatic network configuration, aggregate data collection and graphical configuration.

The module provides complete support during all the steps of the processing chain (see Figure 9.11), from sensor data acquisition up to classification.

The processing chain may involve feedback loops, some of the steps could not be necessary and each single step output might be useful itself for specific dedicated analysis. In particular, *filtering* is typically considered a pre-processing operation which is applied to the raw sensor streams to clean the signals without loosing relevant information. *Stream segmentation* is a complex, but often critical, task because aims at segmenting the signal into segments, each of them containing a certain event to be detected or a phenomenon to recognize. For instance, in the context of physical activity recognition,

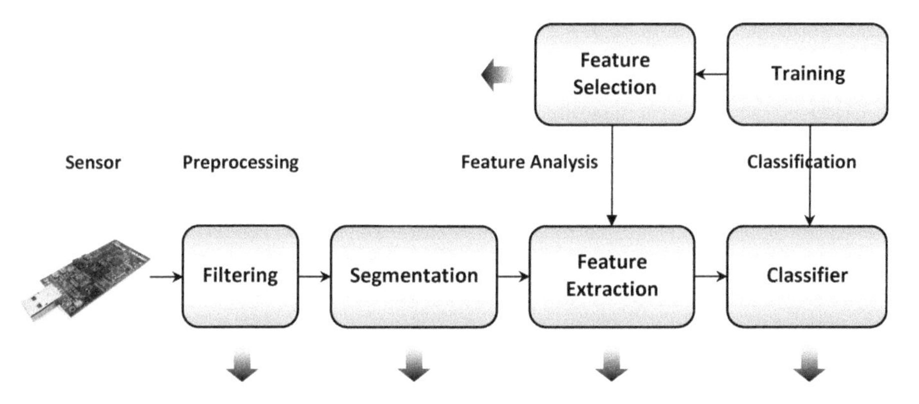

Fig. 9.11 Data processing chain supported by SPINE.

segmentation is performed by splitting the signal into periods of activity and resting, to remove useless data and help identifying portions of the signal which will be later processed to detect specific movements. *Feature extraction* is certainly one of the fundamental steps as it computes specific properties (or features) from the filtered (and possibly segmented) signal to reduce the amount of data available by providing feature sets which characterize the original signal. For instance, most physical activity recognition systems extract time-domain features (such as average value, amplitude, standard deviation) from the motion signals coming from sensors placed on the body. *Feature selection* is a particular step which is necessary only during the first phases of the signal analysis and problem formulation. It determines the optimal subset from the extracted feature set, which means the most significant subset that best characterize the signal for improving the classification accuracy. For instance, a big feature set containing several features from all the axes of each accelerometer or gyroscope sensor placed on the body of an assisted living typically includes several correlated features as human activities often involve full body movements. Once identified the most significant feature set, the feature extraction step is re-configured to compute only the relevant features, and the feature selection task can be removed from the final processing flow. Finally, *classification* uses the feature set to classify new instances by labeling them. There are very different types of classifiers, although for WBSN application, supervised algorithms (such as K-NN, decision trees, or Bayesian) are commonly used. Any supervised algorithm requires a preliminary step called *training* that is necessary to create the specific classifier model for the target application. A training set, containing a set of well-known instances manually labeled, is required for the training phase. Constructing a robust training set is critical to obtain high classification performance as it represents the reference data to which new instances are compared against by the classifier.

9.4 A WBSN-based Monitoring System

The real-time and synchronized monitoring of activity and heart rate of assisted livings, anywhere and anytime, is an important building block not only for e-Health but also for all those application domains that require real-time recognition of human activity status related to the main human vital parameter, which is the heart rate. This integrated information can be used for detection of cardiovascular anomalies, control of fitness activities, recognition of emotional states, analysis of workload stress, feedback to video-games, etc.

In the following we present the development of such a system (see Figure 9.12) through the SPINE framework according to the "single body" "single BS" WBSN architecture. SPINE is a domain-specific open source framework, developed to support the design of optimized WBSN applications while minimizing the design time and effort. SPINE offers a service

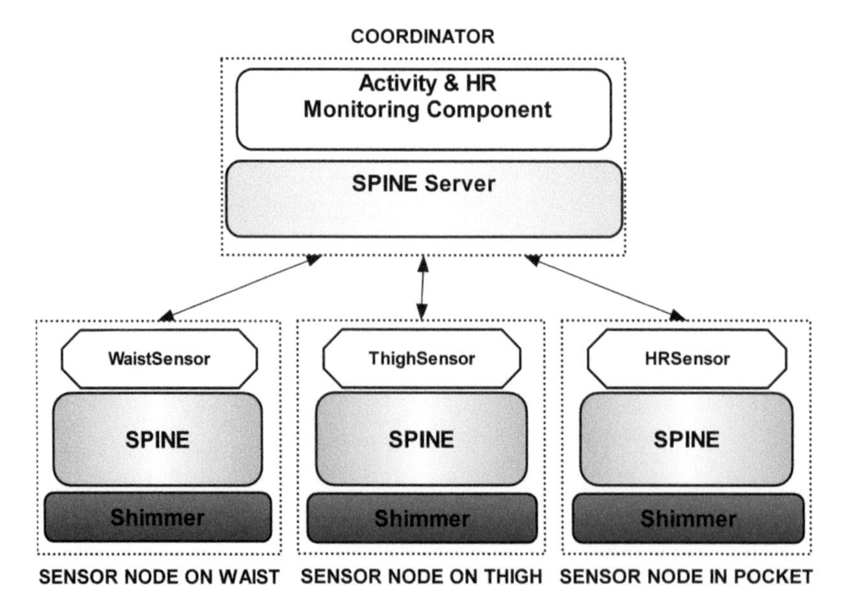

Fig. 9.12 Architecture of the wearable monitoring system.

architecture and a set of libraries that are common to most WBSN applications, and at the same time makes it easy to customize or extend the provided libraries to meet the particular needs of specific applications. SPINE provides libraries of protocols, utilities and processing functions and a lightweight Java API that can be used by local and remote applications to manage the sensor nodes or issue service requests. SPINE supports distributed implementations of classification systems where signal processing functions are computed on the sensor nodes and the result sent to the WBSN coordinator (called SPINE coordinator) running on a PC or smart-phone/PDA. This allows reducing the amount of data exchanged between the WBSN coordinator and the sensor nodes with respect to applications where sensor nodes transmit raw sensor data. The architecture of the developed system at the sensor node side is based on:

- One HRSensor component running on a Shimmer mote [22] interfaced with a Polar band by means of the cardio-shield board [2] to extract the heart rate;
- Two accelerometer components running on Shimmer motes, respectively positioned on the waist (WaistSensor) and the thigh (ThighSensor) of the monitored person to acquire accelerometer data for activity recognition.

HRSensor, upon the reception of the heart pulse from the sensor board, computes the instantaneous heart rate and sends it to the coordinator. The WaistSensor and the ThighSensor sense the 3-axial accelerometer sensor according to a given sampling time and compute specific features on the acquired raw data periodically. Finally, they transmit their results to the coordinator. In particular, the accelerometer on both nodes is sampled at 20Hz, and the following significant features are computed on the nodes over 40-sample windows with a 50% overlap:

- Waist node: mean on the accelerometer axes X and Z, min value and max value on the accelerometer axis X;
- Thigh node: min value on the accelerometer axis X.

The coordinator side is based on the Activity&HR Monitoring (AHRM) component. In particular, the AHRM receives the data from the three above defined sensor components to recognize the current activity of the person and associate his/her heart rate to it in real-time. The AHRM then forwards such information to a higher-level remote monitoring component, located at the control center, which is able to provide real-time information about the status (activity and heart rate) of the group of monitored assisted livings.

9.4.1 Activity Recognition and Heart Rate Calculation

The activity recognition relies on a classifier that takes accelerometer data features (max, min and average; see above) and recognizes the movements defined in a training phase. Among the classification algorithms available in the literature, a K-Nearest Neighbor (KNN)-based classifier [6] has been chosen. The significant features to be activated on the nodes to classify the movements have been selected using an off-line sequential forward floating selection (SFFS) algorithm [17]. Experimental results show that, given a certain training set, the classification accuracy is not much affected by the K value or the type of distance metric used by the classifier. This is because, in this specific example, classes (lying down, sitting, standing still and walking) are rather separate and not affected by noise (see Figure 9.13). Therefore, K set to 1 and the Manhattan distance are used as parameters of the KNN-based classifier.

The average heart rate is computed from the peak-to-peak interval (RRi) (see paragraph 9.3.2).

In particular, the instantaneous rate is calculated directly using the RRi whereas a 20-point moving average is applied to generate a smoother, more robust heart rate diagram.

9.4.2 Graphical User Interface

A snapshot of the AHRM desktop application GUI is shown in Figure 9.14.

The heart rate plot shown in the user interface is dynamically updated in real-time. The activities being performed are displayed on the heart rate diagram using a textual overlay. Note that textual annotations are added only during activity transitions.

A graph, based on data computed by AHRM, which integrates the heart rate with the activity of an assisted living wearing the WBSN-based system,

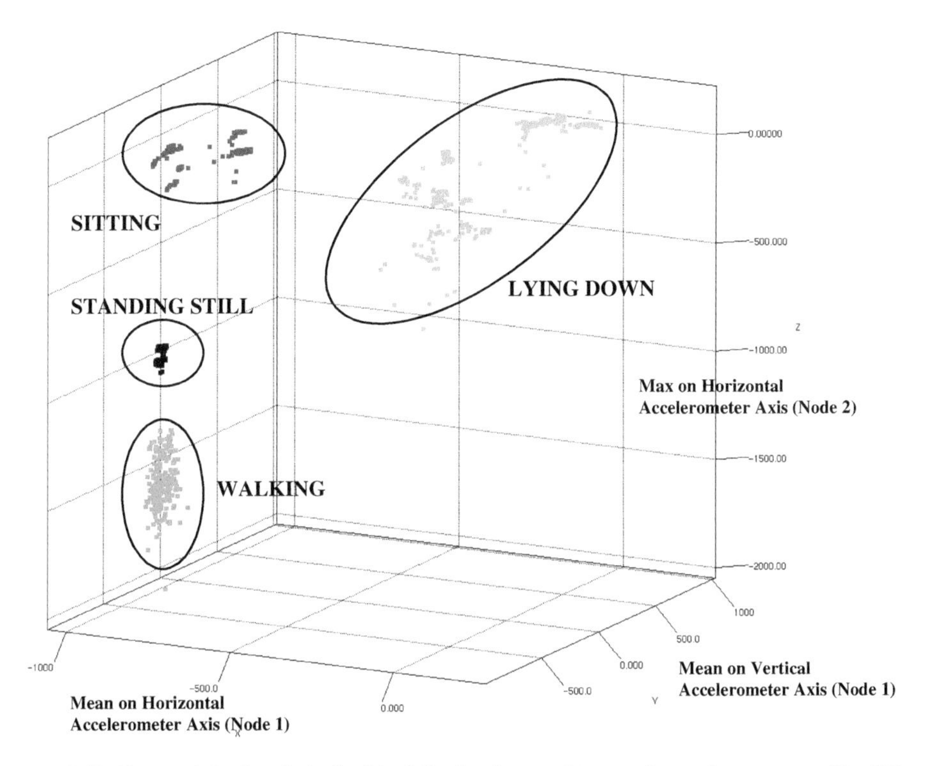

Fig. 9.13 Clusters defined on the basis of the following features: Mean on the accelerometer axes X and Z for the waist sensor node and Max on the accelerometer axis X for the thigh sensor node.

is portrayed in Figure 9.15. The graph shows how the heart rate is affected by the activity of the assisted living.

9.4.3 Data Acquisition and Analysis

The AHRM system can be effective to monitor the heart rhythm during daily life without affecting the patient's comfort, as the sensors involved are wireless and lightweight. In particular, the system can obtain important information by correlating the heart rate with the current activity being performed. For example, it could notify heart rhythms, which are abnormal for the current physical activity, to the patient or directly to his/her doctor in case of alert situations. Furthermore, by analyzing the long-term logged data, early detection of various heart conditions can be achieved, as some initial illness symptoms may not be detected during the short time interval of the visit to the cardiologist office

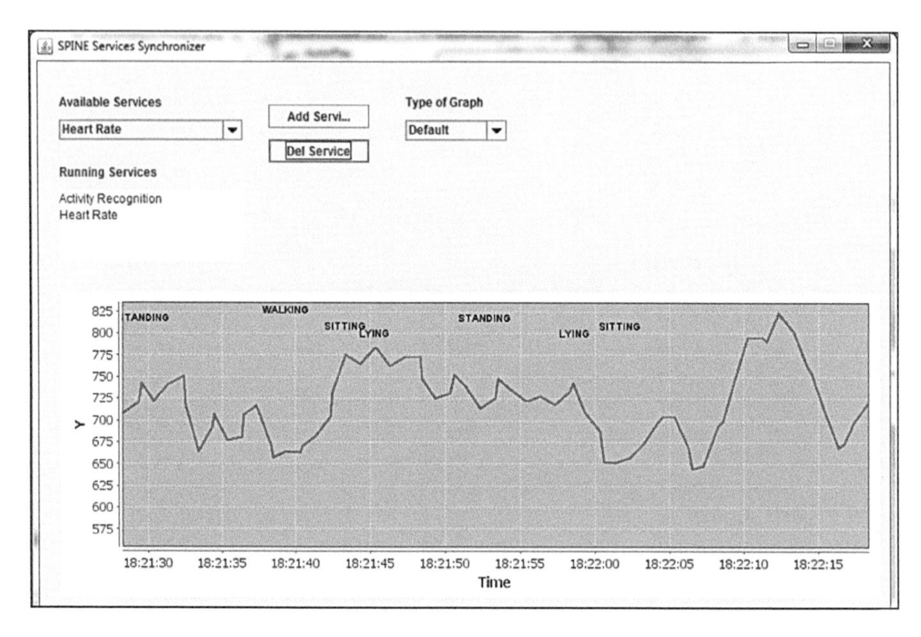

Fig. 9.14 Snapshot of the Activity & Heart Rate Monitoring.

(as some symptoms may be intermittent at an early stage of a heart condition). To provide an idea of the effectiveness of the proposed system, we have monitored for 24 hours two young persons. The first is a male student with a normal heart condition. The second is a person with a diagnosed tachycardia.

Heart-rate diagram shown in Figure 9.16 has been extracted from the young healthy student. By analyzing the overall diagram, we note that the heart-rate rapid increase/decrease is due to the normal body reaction to certain changes of his activity level. Furthermore, by taking a closer look to the heart rate diagram in Figure 9.16, we note the average rate during sitting is slower than the normal 60–100 beats/min (BPM).

This heart condition is known as bradycardia. Bradycardia is the resting heart rate of under 60 BPM, although it is seldom symptomatic until the rate drops below 50 BPM [16]. It is generally considered pathological, although trained athletes or young healthy individuals may also have a slow resting heart rate which, in this case, is absolutely a normal side effect of the training. The AHRM system can easily detect this condition from the joint analysis of the heart rate and the physical activity. Heart rate diagram shown in Figure 9.17 has been extracted from the data collected on the second subject.

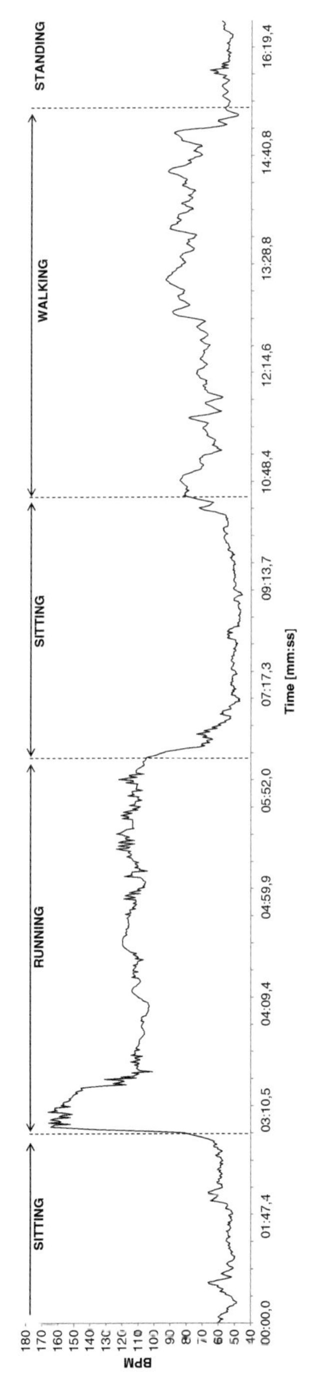

Fig. 9.15 Integrated information about heart rate and activity of an assisted living wearing the WBSN-based system.

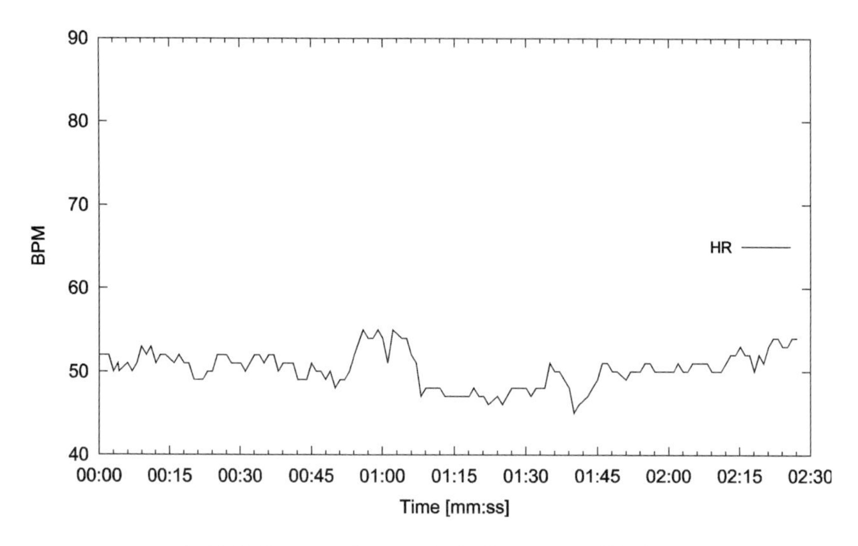

Fig. 9.16 Heart rate diagram of a healthy person while sitting.

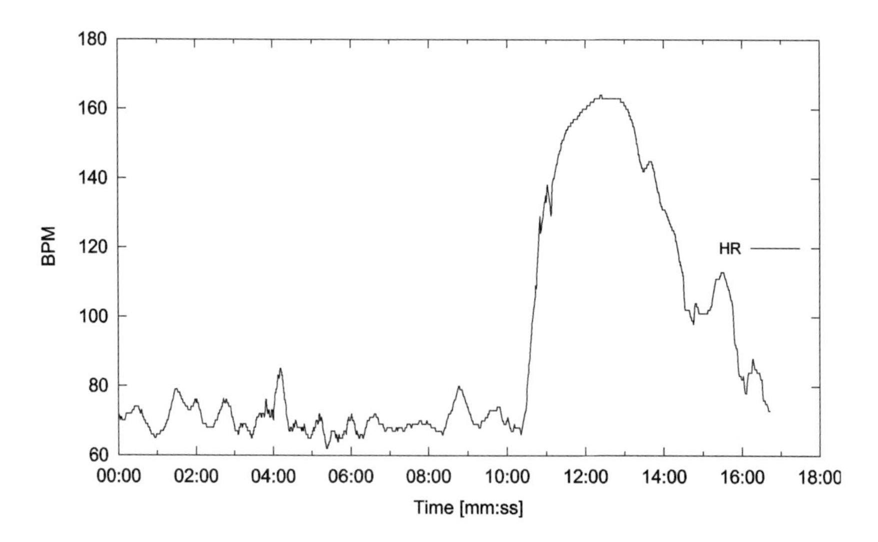

Fig. 9.17 Tachycardic attack while sitting.

It shows a 15 minutes time window in which the subject was seated. At the tenth minute, we note a sudden increase of the heart rate above 160 BPM. The rhythm comes back to normal after about 4 minutes. Although various situations may cause a sudden heart rate increase, because the AHRM system is aware of the fact that in this case the increase is not due to a rapid change of the subject's activity level, it may issue an alarm as this sudden increase could be a symptom of a short tachycardic attack. Tachycardia refers to a heart rate that exceeds the normal range for a resting heart rate. Adults are considered affected of tachycardia when their resting rate is above 100 BPM [16].

These examples have shown cases in which cross-correlating the heart rate with the physical activity is useful to provide indications to both healthy subjects and individuals affected by cardiac diseases. We are currently supported by a cardiologist to enhance the AHRM system with more accurate cross-analysis also taking into account other information relative to the specific person's history that may become important to avoid false alarms (e.g., in the case of the trained athlete for whom the bradycardia is a physiological situation), and to detect possible symptoms of a disease from situations that are instead generally normal (e.g., in certain circumstances, even if the resting heart rate is above 60 BPM, it can be too slow for the individual's current medical condition). For instance, the paroxysmal supra-ventricular tachycardia (SVT) is a rapid rhythm of the heart that involves an accessory pathway. It generates elevated rate of impulses that can reach or even exceed 270 BPM [16]. Symptoms can arise suddenly and may stop without treatment. They are caused for a reason other than stress, exercise, or emotion. They can last a few minutes or as long as a few days. Because wide-complex tachycardia may mimic ventricular tachycardia (VT), it is important to determine whether a wide-complex tachycardia is an SVT or a VT, since they are treated differently. In particular, VT has to be treated appropriately, since it can quickly degenerate to ventricular fibrillation and death. The proposed AHRM system can, therefore, give support to the cardiologist diagnosis, as it is able to show whether the cardiac events are corresponding to physical exercise or emotional stress.

9.5 Conclusions

In this chapter we have proposed WBSNs as enabling technology for a wide variety of novel application scenarios and systems in strategic application domain such as e-Health, e-Emergency, e-Sport, e-Entertainment, e-Factory and e-Sociality. After describing architectures, frameworks and application domains for WBSNs, the chapter described the SPINE Framework and its flexibility for supporting both raw sensor data recording and on-node real-time signal processing functionalities. A case study for monitoring and analysis of the basic activities and heart rate of assisted livings has been also presented. The system could be used in any of the delineated application domains as basic support for more complex tasks performed by the assisted livings. Future work will be devoted to: (i) using the Android SPINE framework for the system implementation atop Android smart-phones/tablets; (ii) the integration of the WBSN system with a cloud-based server-side system (control and analysis center); (iv) an in-depth, cardiologist-supported analysis of the activity/heart rate integrated information in different scenarios of the daily life.

References

[1] P. Alexandros and B. Nikolaos, "A Survey on Wearable Sensor-Based Systems For Health Monitoring and Prognosis," *IEEE Transactions on Systems, Man and Cybernetics*, vol. 40, no. 1, pp. 1–12, 2010.

[2] A. Andreoli, R. Gravina, R. Giannantonio, P. Pierleoni, and G. Fortino, "SPINE-HRV: A BSN-based toolkit for heart rate variability analysis in the time-domain," *Wearable and Autonomous Biomedical Devices and Systems: New issues and Characterization — Lecture Notes on Electrical Engineering*, vol. 75, pp. 369–389, 2010.

[3] A. Augimeri, G. Fortino, M. Rege, V. Handzisky, and A. Wolisz, "A cooperative approach for handshake detection based on body sensor networks," in *Proceedings of the IEEE International Conference on Systems, Man, and Cybernetics*, SMC 2010, pp. 281–288, IEEE Press, October 2010.

[4] F. Bellifemine, G. Fortino, R. Giannantonio, R. Gravina, A. Guerrieri, and M. Sgroi, "Development of body sensor network applications using SPINE," in *Proceedings of the IEEE International Conference on Systems, Man, and Cybernetics*, SMC 2008, pp. 2810–2815, IEEE Press, October 2008.

[5] F. Bellifemine, G. Fortino, R. Giannantonio, R. Gravina, A. Guerrieri, and M. Sgroi, "SPINE: A domain-specific framework for rapid prototyping of WBSN applications," *Software: Practice & Experience*, vol. 41, no. 3, pp. 237–265, March 2011.

[6] T. Cover and P. Hart, "Nearest neighbor pattern classification," *IEEE Transactions on Information Theory*, vol. 13, pp. 21–27, January 1967.

[7] S. Coyle, D. Morris, K. Lau, N. Moyna, and D. Diamond, "Textile-based wearable sensors for assisting sports performance," in *Proceedings of the International Conference on Body Sensor Networks*, BSN 2009, pp. 228–233, IEEE Computer Society, June 2010.

[8] R. Gravina, A. Andreoli, A. Salmeri, L. Buondonno, N. Raveendranathank, V. Loseu, R. Giannantonio, E. Seto, and G. Fortino, "Enabling multiple BSN applications using the SPINE framework," in *Proceedings of the International Conference on Body Sensor Networks*, BSN 2010, pp. 228–233, IEEE Computer Society, June 2010.

[9] B. Grundlehner, L. Brown, J. Penders, and G. Gyselinckx, "The design and analysis of a real-time, continuous arousal monitor," in *Proceedings of the 6th International Workshop on Wearable and Implantable Body Sensor Networks*, BSN 2009, pp. 156–161, IEEE Press, June 2009.

[10] Y. Hao and R. Foster, "Wireless body sensor networks for health-monitoring applications," *Physiological Measurement*, vol. 29, no. 11, R27–R56, November 2009.

[11] G. Holmes, A. Donkin, and I. H. Witten, "Weka: A machine learning workbench," in *Proceedings of the 2nd Australia and New Zealand Conference on Intelligent Information Systems*, ANZIIS'94, pp. 1269–1277, IEEE Press, 1994.

[12] J.-Y. Huang and C.-H. Tsai, "A wearable computing environment for the security of a large-scale factory," in *Proceedings of the 12th International Conference on Human-computer interaction: interaction platforms and techniques*, HCI'07, pp. 1113–1122, Springer-Verlag, July 2007.

[13] C. Lombriser, D. Roggen, M. Stager, and G. Troster, "Titan: A tiny task network for dynamically reconfigurable heterogeneous sensor networks," in *Proceedings of the 15th Fachtagung Kommunikation in Verteilten Systemen*, KiVS 2007, pp. 127–138, Springer, February 2007.

[14] K. Lorincz, D.-J. Malan, T. Fulford-Jones, A. Nawoj, A. Clavel, V. Shnayder, G. Mainland, M. Welsh, and S. Moulton, "Sensor networks for emergency response: Challenges and opportunities," *IEEE Pervasive Computing*, vol. 3, no. 4, pp. 16–23, October 2004.

[15] C.-W. Mundt, K.-N. Montgomery, U.-E. Udoh, V.-N. Barker, G.-C. Thonier, A.-M. Tellier, R.-D. Ricks, R.-B. Darling, Y.-D. Cagle, N.-A. Cabrol, S.-J. Ruoss, J.-L. Swain, J.-W. Hines, and G.-T. Kovacs, "A multiparameter wearable physiologic monitoring system for space and terrestrial applications," *IEEE Transactions on Information Technology in Biomedicine*, vol. 9, no. 3, pp. 382–391, September 2005.

[16] J. Olgin and D. Zipes, "Specific Arrhythmias: Diagnosis and treatment," in *Braunwald's Heart Disease: A Textbook of Cardiovascular Medicine, 8th ed.*, pp. 863–931, Philadelphia: Saunders Elsevier, 2008.

[17] P. Pudil, J. Novovicova, and J. Kittler, "Floating search methods in feature selection," *Pattern Recognition Letters*, vol. 15, no. 11, pp. 1119–1125, November 1994.

[18] N. Raveendranathan, S. Galzarano, V. Loseu, R. Gravina, R. Giannantonio, M. Sgroi, R. Jafari, and G. Fortino, "From modeling to implementation of virtual sensors in body sensor networks," *IEEE Sensors Journal*, doi:10.1109/JSEN.2011.2121059, 2011.

[19] X. F. Teng and Y. T. Zhang, "Continuous and noninvasive estimation of arterial blood pressure using a hotoplethysmographic approach," in *Proceedings of the 25th Annual International Conference of the IEEE Engineering in Medicine and Biology Society*, EMBS'03, pp. 3153–3156, IEEE, September 2003.

[20] T. Terada and K. Tanaka, "A framework for constructing entertainment contents using flash and wearable sensors," in *Proceedings of the 9th International Conference on Entertainment computing*, ICEC'10, pp. 334–341, Springer-Verlag, 2010.

[21] Fitbit website. http://www.fitbit.com, 2011.

[22] Shimmer website. http://www.shimmer-research.com/, 2011.

[23] SPINE website. http://spine.tilab.com, 2011.

[24] SunSPOT website. http://www.sunspotworld.com, 2011.

[25] TinyOS website. http://www.tinyos.net, 2011.

[26] Vitalsense website. http://vitalsense.respironics.com, 2011.

[27] F. Yarrow and J. Rice, "Localized IR spectroscopy of hemoglobin," *European Biophysics Journal*, vol. 40, no. 2, pp. 217–219, September 2010.

[28] J. Yick, B. Mukherjee, and D. Ghosal, "Wireless sensor network survey," *Computer Networks*, vol. 52, pp. 2292–2330, 2008.

[29] M. Zhang and A. Sawchuk, "A customizable framework of body area sensor network for rehabilitation," in *Proceedings of the 2nd International Symposium on Applied Sciences in Biomedical and Communication Technologies*, ISABEL 2009, pp. 24–27, IEEE Press, November 2009.

[30] D. Zikich, "Arterial blood flow sensor," in *Proceedings of the 4th European Conference of the International Federation for Medical and Biological Engineering*, EMBEC'09, pp. 1158–1162, 2009.

Index